War Time

The International Society for First World War Studies' ninth conference, 'War Time', drew together emerging and leading scholars to discuss, reflect upon, and reconsider the ways that time has been conceptualised both during the war itself and in subsequent scholarship. *War Time: First World War Perspectives on Temporality*, stemming from this 2016 conference, offers its readers a collection of the conference's most inspiring and thought-provoking papers from the next generation of First World War scholars. In its varied yet thematically related chapters, the book aims to examine new chronologies of the Great War and bring together its military and social history. Its cohesive theme creates opportunities to find common ground and connections between these sub-disciplines of history, and prompts students and academics alike to seriously consider time as alternately a unifying, divisive, and ultimately shaping force in the conflict and its historiography. With content spanning land and air, the home and fighting fronts, multiple nations, and stretching to both pre-1914 and post-1918, these ten chapters by emerging researchers (plus an introductory chapter by the conference organisers, and a foreword by John Horne) offer an irreplaceable and invaluable snapshot of how the next generation of First World War scholars from eight countries were innovatively conceptualising the conflict and its legacy at the midpoint of its centenary.

Louis Halewood is a DPhil student in History at Merton College, University of Oxford. His research analyses visions of a new world order, and the role of maritime power in its creation and underpinning, in Britain, France, and the United States between 1890 and 1922.

Adam Luptak is a DPhil student in History at Oriel College, University of Oxford. His research explores the topic of disabled veterans of the Great War in interbellum Czechoslovakia.

Hanna Smyth is a DPhil student in Global and Imperial History at Exeter College, University of Oxford. Her research is transnational comparison of Imperial War Graves Commission sites on the Western Front, examining how they represented different aspects of South African, Indian, Canadian, and Australian identities in the 1920s–1930s.

Routledge Studies in First World War History

Series editor
John Bourne
The University of Birmingham, UK

The First World War is a subject of perennial interest to historians and is often regarded as a watershed event, marking the end of the nineteenth century and the beginning of the 'modern' industrial world. The sheer scale of the conflict and massive loss of life means that it is constantly being assessed and reassessed to examine its lasting military, political, sociological, industrial, cultural and economic impact. Reflecting the latest international scholarly research, the Routledge Studies in First World War History series provides a unique platform for the publication of monographs on all aspects of the Great War. Whilst the main thrust of the series is on the military aspects of the conflict, other related areas (including cultural, visual, literary, political and social) are also addressed. Books published are aimed primarily at a post-graduate academic audience, furthering exciting recent interpretations of the war, whilst still being accessible enough to appeal to a wider audience of educated lay readers.

Also in this series

Military Service Tribunals and Boards in the Great War Determining the Fate of Britain's and New Zealand's Conscripts
David Littlewood

The Royal Flying Corps, the Western Front and the Control of the Air, 1914–1918
James Pugh

The Great War and the British Empire Culture and Society
Edited by Michael J.K. Walsh and Andrekos Varnava

Aerial Propaganda and the Wartime Occupation of France, 1914–18
Bernard Wilkin

For more information about this series, please visit: https://www.routledge.com/history/series/WWI

War Time

First World War Perspectives on Temporality

Edited by
Louis Halewood, Adam Luptak, and Hanna Smyth

Routledge
Taylor & Francis Group
LONDON AND NEW YORK

First published 2019
by Routledge

2 Park Square, Milton Park, Abingdon, Oxfordshire OX14 4RN
52 Vanderbilt Avenue, New York, NY 10017

Routledge is an imprint of the Taylor & Francis Group, an informa business

First issued in paperback 2020

British Library Cataloguing in Publication Data
A catalogue record for this book is available from the British Library

Library of Congress Cataloging in Publication Data
A catalog record for this book has been requested

ISBN: 978-1-138-30835-0 (hbk)
ISBN: 978-0-367-58892-2 (pbk)

Typeset in Times New Roman
by Taylor & Francis Books

Contents

Foreword

Time, the very medium of history, has all too rarely been made the object of historical enquiry and of nothing is this truer than the First World War, as the editors of this book observe. Yet that episode, the first great upheaval of the twentieth century and so of the contemporary world, shook time itself.

It did so, first, because contemporaries were acutely aware that they were living through a moment of epochal change. By the measure of historical time, no event since the French Revolution and the Napoleonic Wars had, for Europeans at least, assumed such epic proportions and seemed so certain to change the destinies of countries, empires, and regimes. Just how true this was only time would tell, but it did so in ways that no one predicted when the war broke out. Each twist in the long-term legacies of the Great War rewrote the potential verdict of time, the junctures and periods into which the event seemed to divide history. If 1918 was reversed for France and Germany in 1940 and again in 1945, the dialectic that Communists saw in the November 1917 revolution was cancelled in 1989–1991 by the fall of the Soviet Union. This now made the war the immediate source of the greatest failed utopia of modernity. The temporal genealogies of colonialism and anti-colonialism in relation to the war are still being worked out.

Second, time (in various registers) was integral to the war itself. It was an organisational and operational factor of the first importance. Mobilising armies, coordinating attacks, establishing logistical flows across continents and oceans all had the effect of generalising standard time, globally agreed as recently as 1884, and inculcating wide familiarity with the 24-hour military clock. Trench warfare turned wristwatches into the universal timekeeping device they remained until the advent of the mobile phone, supplanting the unwieldy pocket watch of the later nineteenth century. Yet internalised standard time stood in sharp contrast to subjective time, something on which philosophers like Bergson and the great avatars of modernism wrote extensively in the decades spanning the conflict (Proust completed his novel cycle *In Search of Time Past* not only during the war but by incorporating the war into his text). Personal time was elastic, stretching from an eternal present of waiting to the compressed instant of an air raid or attack. Its grammar included a past tense of sharpened memories and a future so uncertain that for tens of millions of soldiers and their families it dissolved into multiple uses of the conditional tense, turning on death, mutilation or disappearance, and survival or return.

Third, as the editors also note, the inscription of the war into larger timeframes involved more than retrospective judgements as to its causes and consequences, with adjustments to the periodic tables that are central to the historian's craft. Individuals and groups could not fail to assign the war a major place in their memories, something that (like its legacies) changed over time. And as they also note, the centenary itself has been a major point of recalibration, measuring what the war now seems to mean at a century's distance. All of the above (and more) are the subject matter of *War Time*, and the editors of the volume like its contributors are to be commended for taking on this crucial subject.

But, then, such intellectual initiative is nothing new in the International Society for First World War Studies (and its journal *First World War Studies*). For the Society was established in 2001, 15 years before the conference that generated this volume was held, at the mid-point of the centenary in November 2016. It has published a series of conference proceedings, each of which has acted as a showcase for the doctoral and early career scholars who established the society and who then passed on its ambition to drive the field forward to succeeding generations. However, only one university, Oxford, has hosted a conference twice. This fact owes everything to the intersection of the Society's goals with another source of intellectual energy: the collaborative research programme of the Oxford History Faculty, 'Globalising and Localising the Great War', led by Adrian Gregory and administered by Jeanette Atkinson. This programme has federated the energies of a stellar group of doctoral students into an academic creativity greater even than the sum of its parts.

The editors (and it speaks well of them) are honest and self-critical as to the limits of the book, regretting, for example, the relative absence of chapters that go beyond the Western Front. What time and experience teach, of course, is that all such books are (or should be) limited by the measure of their ambition. That is the price paid for their achievement, which is to show the way forward in a new direction. Inverting the bottle may leave it as half-empty (or half-full) as before, but it should point to what has been achieved. That is a series of essays, based on new and exciting research, which show just how pertinent the theme of time is for the study of war. They demonstrate, if need there was, that 'war time' is a multi-angled tool for revealing new faces of human conflict and that in this, as in so much else, the First World War is an ideal observatory and workshop. Above all, they show that at its centenary, the Great War remains one of the richest and most exciting sites of scholarship for the contemporary period and that, in the hands of a rising generation of historians, it is set to remain so for the foreseeable future.

John Horne
Dublin, January 2018

Contributors

Alina Enzensberger is a PhD student at Humboldt University of Berlin. She studied history, philosophy, and social sciences at Humboldt University and University College London. Her dissertation examines "Military hospitals at the German home front as transitory spaces in the Great War, 1914–1918" and is supervised by Professor Gabriele Metzler. This research has been supported by a scholarship of the Franco-German University (DFH/UFA) and by the 2016 award of the International Centre for Research of the Historial de la Grande Guerre. Currently, Alina Enzensberger is a fellowship holder at the Centre Marc Bloch (CMB) in Berlin.

Ashley Garber is completing her DPhil in history at Oxford University and holds an MA with distinction in war studies from King's College London. Her work examines memory discourses within ex-servicemen's organisations in Britain and America during the twentieth century. She is especially interested in comparative and interdisciplinary approaches to the history of war and memory.

Anna Grillini obtained her PhD at the University of Trento in 2016 and is currently a post-doc researcher at the German Historical Institute in Rome. Her studies focus on the history of psychiatry and trauma in soldiers and civilian population during the Great War. Among her publications is *La guerra in testa: Esperienze di civili, profughi e soldati nel manicomio di Pergine Vasugana (1909–1924)*, Annali dell'Istituto Storico Italo-Germanico, in press, 2018.

Erica Grossi lives in Milan. After graduating in Contemporary History (University of Pisa – EPHE, Paris), she earned a PhD in European Cultural Studies (University of Palermo – Heinrich-Heine-Universität, Düsseldorf) with a thesis on the Italian photo albums of the First World War and the visual culture between the wars. She collaborates with Italian research institutions (Fondazione Feltrinelli and Istituto Parri, Milan; Archivio della Scrittua Popolare, Genoa). Among her recent publications are "Naufragio Apocalisse Malattia: Metafore dell'esistenza nel secolo della Grande Guerra", in S. Ballerio, E. Belloni, and E. Grossi, *I percorsi* *della Grande Trasformazione* (Quaderni/16, Fondazione Giangiacomo Feltrinelli, 2017), 55–94: http://fondazionefeltrinelli.it/schede/ebook-percorsi-della-grande-trasformazione-9788868352745/); *Walter Benjamin: Arte, media, filosofia della storia. Per un'archeologia dei tempi moderni* (Hachette, 2016).

Louis Halewood is a DPhil student in History at Merton College, University of Oxford. His research analyses visions of a new world order, and the role of maritime power in its creation and underpinning, in Britain, France, and the United States between 1890 and 1922.

Sante Lesti is a Postdoctoral Research Fellow in Modern European History at the Scuola Normale Superiore (Pisa). He obtained a joint PhD degree in History and Civilisation from the Scuola Normale Superiore (Pisa) and the École des Hautes Études en Sciences Sociales (Paris). His main research interests are religion and politics and the cultural history of the First World War. In 2015 he published his first monograph, *Riti di guerra. Religione e politica nell'Europa della Grande Guerra* (Bologna: il Mulino).

Adam Luptak is a DPhil student in History at Oriel College, University of Oxford. His research explores the topic of disabled veterans of the Great War in interbellum Czechoslovakia.

Alex Mayhew is a final year PhD candidate and scholarship holder in the Department of International History at the London School of Economics and Political Science, where he also teaches social sciences. He works under the supervision of Dr Heather Jones and Professor David Stevenson. His thesis, "Making Sense of the Western Front: English Infantrymen's Morale and Perception of Crisis during the Great War", focuses on the relationship between soldiers' environment, their perception of crisis, and their morale. In 2017 he won a Gerda Henkel scholarship from the Centre International de Recherche de l'Historial de la Grande Guerre, Péronne.

Steve Marti received his PhD from the University of Western Ontario in 2015. His dissertation examines how categories of race, class, and gender shaped the relationship between voluntary mobilisation and community in Australia, Canada, and New Zealand. He has published articles in *Histoire sociale/Social History, Itinerario, History Australia*, and *The Journal of Australian Studies.* Steve recently completed a postdoctoral fellowship in digital history at the University of Delaware and continues to work as an independent scholar.

Assaf Mond is a PhD student at the Zvi Yavetz School of Historical Studies in Tel Aviv University. He is writing his PhD thesis, "The Changing Urban Space of Great-War London, 1914–1918", under the supervision of Prof. Iris Rachamimov. His article "Chelsea Football Club and the Fight for Professional Football in First World War London" was published in *The London Journal* in November 2016.

Robert Rennie is a historian of modern Germany and specialises in the history of technology. He is a graduate of the University of North Georgia, Appalachian State University, and he earned his PhD from the University of Tennessee, Knoxville in 2017. Rennie's work in the classroom has been recognised with several outstanding teaching awards while his research recently earned him the John H. Morrow Award for Excellence in Military History. He has presented his research at the International Society for First World War Studies, the Council for European Studies, and the Southeast German Studies Workshop.

Hanna Smyth is a PhD candidate in Global and Imperial History at Exeter College, University of Oxford. Her research is transnational comparison of Imperial War Graves Commission sites on the Western Front, examining how they represented different aspects of South African, Indian, Canadian, and Australian identities in the 1920s–1930s.

Cedric Van Dijck is reading for a PhD in English Literature at Ghent University (FWO). His research on modernism and the trench press has appeared or is forthcoming in *Modernist Cultures, Publishing History, TSLL*, and *PMLA*, and has received further recognition from the Historial de la Grande Guerre and the British Association for Modernist Studies. He recently co-edited *The Intellectual Response to the First World War* (2017).

Acknowledgements

This volume is a selected edition of the proceedings of the ninth conference of the International Society for First World War Studies, 'War Time', with ten of the conference's 18 papers published here in expanded and revised form. 'War Time' took place at the University of Oxford on 10–11 November 2016. The editors and the other seven members of their conference team became involved in organising 'War Time' through the 'Globalising and Localising the Great War' network at Oxford, to which they all belonged as PhD students or postdoctoral researchers.

The support of members of both organisations was vital to the success of the conference and this publication, as was the enthusiasm and support of those who presented, chaired, and gave commentaries and keynotes at the conference. Thus, in addition to the authors in this volume, we would particularly like to thank Emily Anderson, Jeanette Atkinson, Charlotte Bennett, Jonathan Boff, Laurence Campa, Selena Daly, Ellen Davies, Jack Doyle, Marc-André Dufour, Aimée Fox, Tim Godden, Adrian Gregory, Franziska Heimburger, John Horne, Patrick Houlihan, Greg Hynes, Michael Joseph, Alice Kelly, Kate Kennedy, Jonathan Krause, Jenny Macleod, Margaret MacMillan, Helen McCartney, Christoph Mick, Claire Morelon, David Morgan-Owen, Michael Neiberg, John Paul Newman, Jane Potter, Catriona Pennell, Heather Perry, Pierre Purseigle, Frank Reichherzer, Felix Schmidt, Jan Schmidt, Steffen Rimner, Nicholas Rodger, Martina Salvante, Thomas Schmutz, Sir Hew Strachan, Calum White, and Vanda Wilcox, as well as all others who attended.

Further thanks are due to the Maison Française d'Oxford, who generously hosted the conference. We also very much appreciate the 22 academics who served as peer reviewers for this volume (each chapter was double-peer-reviewed). At Routledge, we are very grateful to Rob Langham, Julie Fitzsimons, Seth Townley, and freelance copyeditor Neil Dowden, and to the anonymous reviewers of our manuscript proposal.

Introduction

Louis Halewood, Adam Luptak, and Hanna Smyth

Time serves as a potentially invaluable category and tool of scholarly enquiry in First World War studies. The subfield of First World War memory studies implicitly deals with concepts and impacts of time: to give just a few examples, West and Aarons have addressed generational war memory in Australia; Scates has written on the intersection of 'war time' and 'war place' through analysing contemporary revisitation of battlefield landscapes; and Sear has explored how the changing temporalities of the digital age have impacted the transmission of First World War memory.[1] However, few works have explored the relationship of war and time *explicitly* and *substantially* in First World War contexts. As a result of this general absence of scholarship, many important questions remain unanswered.

Time has always been an integral part of human life and experience. As Daniel Rosenberg and Anthony Grafton's study *Cartographies of Time* aptly illustrates, mankind has been intrigued by time and timelines for many centuries.[2] Similarly, people have always been fascinated by armed conflict for a plethora of different reasons, and throughout its existence mankind has spent an extraordinary amount of time and energy remembering (or trying to forget), discussing, and analysing war. Although the war cannot be compared with time in terms of its absolute and constant omnipresence, it is perhaps not unfair to say that war and armed conflict in general have been one of only very few subjects to have successfully competed with time for human interest and importance.

Despite the salient role of war and time in history, only a small number of works in existing Anglophone historiography have examined connections between temporality and war. Cheryl A. Wells's *Civil War Time: Temporality and Identity in America, 1861–1865* offers a broad exploration of the impact of war on how groups and individuals perceived time both on and away from the battlefields of the American Civil War.[3] Paul Fussell's book *Wartime* is subtitled *Understanding and Behaviour in the Second World War*, and examines perceptions versus realities of the speed and nature of warfare during that conflict.[4] More recently, Mary L. Dudziak has published *War Time: An Idea, Its History, Its Consequences.* [5] Dudziak explores the theory of 'war time' and how contemporary understanding of this term affects law and politics in the

United States. This work charts the relationship between time and conflict through both the Second World War and the Cold War, although its primary purpose is to relate the concept of periodisation to the development of American politics and society in the aftermath of 9/11 and the Global War on Terror. The use of temporality in the discursive framing of the Global War on Terror by the George W. Bush administration also has received attention from a political science perspective via Lee Jarvis, who explores how American leaders used temporality to narrate the development of their response to 9/11.[6]

Not surprisingly, 'time' has been the subject of considerable scholarly attention in recent years. Amongst these, the works of the late Reinhart Koselleck and the French historian François Hartog stand out as particularly important. Koselleck's studies, which discuss and examine natural chronology and the plurality of historical times,[7] as well as Hartog's analysis of "regimes of historicity", have proved to be particularly influential.[8]

These works have provided important methodological and definitional foundations for future explorations of 'war time', while remaining narrow in focus. These analyses are confined to American perspectives on wars involving the United States, and the First World War occupies a notable gap between the conclusions of Wells's study and the opening of Dudziak's research. Given the subsidiary role traditionally played by the First World War in American narratives, this is perhaps unsurprising. Yet the deluge of new publications on the war – including the role of the United States – at its centenary has not prompted any major studies on the relationship between the war and temporality.[9]

Among the existing works on conflict and temporality, few apply the combination of these themes specifically to the First World War. The historiography of the First World War has expanded significantly in the previous few decades, with many old assumptions challenged and interdisciplinary perspectives introduced.[10] Yet time has been neglected among these, so this volume opens new angles from which to approach it. Crucially, how people *experienced* time during the war and its aftermath; how they were *affected* by the changing perceptions and manipulations of time; how they *utilised* time to convey specific messages; and how their *understandings* of time changed due to the conflict are underexplored avenues which this volume brings to light.

'Time' and the First World War is a hitherto underexplored subfield. However, a notable exception to this is Stephen Kern's *The Culture of Time and Space, 1880–1918*, which addresses the relationship between the First World War and time and how this relates to 'modernity'.[11] Other works interrogating the concept of 'time' less explicitly, but by engaging in the debate concerning modernism and the First World War are implicitly doing so, include Eksteins, Winter, Hynes, and Fussell.[12]

Sister subfields of 'time and the First World War', those of the periodisation and commemoration of the conflict, have received substantially more critical attention during the centenary of this conflict. Unsurprisingly, the periodisation of the Great War has been discussed on many occasions in recent years. Various studies repeatedly point out that the conflict should not be confined to events taking place within the boundaries of the years 1914–1918, but instead should be seen within the context of a much larger timeframe. In his fascinating book *The Vanquished: Why the First World War Failed to End, 1917–1923,* Robert Gerwarth argues that in late 1918 the conflict was far from over in Central and Eastern Europe.[13] Instead, he observes how violence continued into the post-war years, often being directed towards members of ethnic and racial minorities. Taking an approach focusing on the multiplicity and protractedness of the war's end, the edited volume *The Long End of the First World War: Ruptures, Continuities and Memories* posits that the war can be conceptualised as having several 'endings' spanning multiple years after 1918.[14] Conversely, John Horne has argued that there were multiple 'waves' of First World War beginnings, spanning at least three years.[15] Geppert, Mulligan, and Rose, meanwhile, look slightly further back to locate the beginnings of the First World War in various prior conflicts from 1900 to 1914.[16]

Although for any country the period 1914–1918 cannot be understood within a contextual vacuum, for some countries and regions – including particularly Ireland, Russia, the Balkans, and the Ottoman Empire – 1914–1918 is not the most relevant or applicable timeframe with which to understand their relationships with the war. Nevertheless, despite this changing understanding of the conflict and its duration, much of the recent centenary-related attention has echoed the 'traditional' perception of the Great War, memorably delimited by the bullets of Gavrilo Princip on the one side and the armistice of 11 November 1918 on the other.

This edited volume stems from the 'War Time' conference of the International Society for First World War Studies held in Oxford in November 2016. The ISFWWS has been an instrumental institution in the field of First World War studies since its founding in 2001, and actively encourages and promotes the work of emerging scholars. This volume brings together essays written by PhD students and early career researchers covering a broad geographic and temporal range. This allows the volume to offer an international and transnational outlook on the First World War and temporality, produced by a new generation of scholars who are presenting some of the most current research in the field. Their work here addresses questions of both subjective and organisational 'war time' temporalities: organisational because military operations and societal mobilisation required 'modern' time on an unprecedented scale, and subjective because the war multiplied and fragmented personal 'temporalities'.[17]

The title of the conference and this eponymous book has been chosen to reflect the complex and multi-faceted relationship between war and time. Whereas the term 'wartime' has been widely understood as a circumscribed period and state of existence denoting that a declared war is currently being fought, the expression 'war time' has considerably broader connotations; 'war time' allows a wider and richer understanding of the way these two concepts intersect. The essays in this book discuss a variety of themes, which are brought together by their shared position at this intersection.

It is no coincidence that this volume appears roughly one hundred years after the First World War. The centenary has stimulated a veritable tidal wave of both public attention and scholarly research on the war, its memory, and its legacies. The mass collection and digitisation project *Europeana14–18* has made the material culture of transnational and individual memory widely accessible. Public art projects and living memorials have captured widespread and interest, including the 'Tower Poppies' ("Bloodswept Lands and Seas of Red", Paul Cummins and Tom Piper, 2014) and the "We're Here Because We're Here" mass soldier re-enactor installation (Jeremy Deller, 2016, for 14–18 NOW, an organisation that has commissioned numerous First World War centenary art projects).[18]

Scholars have taken the opportunity of the centenary to turn a critical eye towards centenary commemoration itself – such as Beaumont's five-way transnational comparison of the politics of centenary memory, and Catriona Pennell and Sir Hew Strachan's examinations of academics' responsibilities during the centenary[19] – yet they have also found the centenary a prompt for examining themes of interpretation and transmission with new perspectives. By the time the centenary began, direct living memory of the conflict had died out. This has created an increased emphasis on the importance of generations as a periodisation framework for the war (as Ashley Garber's chapter in this volume, "Brothers – and Sons – in Arms", also explores, in a different period), and on the immense power that educators and museums wield in shaping the First World War knowledge of new, younger generations. The former is the focus of ongoing research by Pennell,[20] and in the latter category, it is arguably the Imperial War Museum that has received the most scrutiny, with publications exploring its interpretation and responsibilities from both staff and academic perspectives.[21] Centenary scholarship and public engagement, despite its wide range of manifestations, approaches, and messages, has one universal theme: the relationship between past, present, and future. This relationship, as Erica Grossi's and Sante Lesti's chapters in this volume ("The Photo Albums of the First World War" and "Between Passatism and Futurism", respectively) also argue, demonstrates that the passing of time – more typically associated with forgetting and the growth of temporal distance – can both foment new and reinvigorate old curiosities.

The field of First World War memory studies, and specifically its subfields of periodisation and centenary reflections, are not the only existing historiographical debates this volume engages with. Another major controversy which continues to rage concerns strategies employed on the Western Front. In the aftermath of the war, policymakers and strategists attempted to apportion blame for the heavy casualties suffered during the fighting. In Britain, politicians sought to exculpate themselves by pointing the finger of blame at the generals, creating what Brian Bond has termed "the political myth".[22] In their memoirs, former Prime Minister David Lloyd George (assisted by the renowned military theorist Basil Liddell Hart) and Winston Churchill claimed that the generals had been stubborn, deluded, and unimaginative, resulting in needless losses.[23] These views were entrenched further in the 1960s by the likes of Alan Clark and underscored by Joan Littlewood's theatrical production of *Oh, What a Lovely War!* [24] Counter-arguments that the generals were not so callous and understood their strategy, put forward by the likes of John Terraine, made little headway.[25] New methodological approaches in the 1970s – notably histories of the fighting 'from below' and postmodernist views as presented in Paul Fussell's *The Great War and Modern Memory* – struck the same chord that the war was futile and unnecessary.[26]

In recent years, historians such as Gary Sheffield and William Philpott have made progress in eroding this view.[27] Building on Terraine's analysis, they have explained that the strategy of attrition employed was a valid response to the challenge posed by the deadlock on the Western Front. Given the lack of viable alternatives, it was accepted by contemporaries as necessary. The devastation and slaughter was not senseless, but driven by the logic of attrition. Yet many questions remain within this interpretation. For instance, how did strategists gauge the success of a strategy of attrition?[28] How did temporality interact with the formulation and implementation of strategy?

Alina Enzensberger's and Anna Grillini's chapters ("No Time To Waste" and "Fast Therapy and Fast Recovery", respectively) break new ground in further opening up this area of debate. They examine the role of temporal pressures in the strategies of attrition pursued by the major belligerents. Moving away from the traditional emphasis on speed in studies of the outbreak of the war, epitomised by A.J.P. Taylor's *War by Timetable*, they focus on the speed with which the various armies were able to return wounded men to combat so to keep their manpower levels stable.[29] The essays suggest that while previous studies are correct to argue that the attrition of men and materiel was critical to winning the war, these were not singular instances; rather they mattered across time too. The speed with which servicemen could be (re)deployed was of vital importance in attritional warfare. Their findings prompt larger questions for historians to grapple with in terms of the overarching conduct of the war at the grand strategic level. Additional avenues to explore in relation to temporality may include the speed of recruitment and mobilisation, or the pace of industrial mobilisation. Whereas historians previously have emphasised material staying power in a long, drawn-out conflict, Enzensberger and Grillini demonstrate that in a static war of attrition time was just as imperative as in a war of manoeuvre where speed decides battles, such as the Battle of France (1940).

Time is often seen as a constant, a fundamental and reliable indicator and measurement of experience. Exploring ways in which this was *not* the case for the people of the First World War allows for a greater and richer understanding of the multiplicity of experiences it produced, and also highlights the ways in which life during the conflict and its aftermath diverged substantially from the contours of our own modern-day experience.

As the volume demonstrates, time permeated every aspect of war experience, and the various chapters in this volume seek to address it in a number of different contexts. Emerging scholars explore the topics of war time and home, fighting, and healing fronts, as well as strategy, religion, memory, medicine, civilians, and hope in a number of different settings. Some chapters focus on individual states whilst others offer comparative or transnational approaches.

The volume in total encompasses ten chapters which have been thematically divided into three sections. In Part I, four authors explore war time through the lens of speed and tempo. Alina Enzensberger's chapter ("No Time to Waste") explores the temporal dynamics within German home-front hospitals during the war. She explains how, following the failure of the 'Schlieffen Plan', Germany found itself in a race against time to win a war against a better-resourced and numerically superior enemy. She explores the inverse of recent historiographical arguments about the strategy of attrition pursued by the major combatants, moving the focus away from the reduction of enemy soldiers to how to return wounded men to the front as quickly as possible. The German War Office feared that recovery times in home-front hospitals were too long, and sought to pressure medical professionals into speeding up patients' discharge. In turn, standards for fitness were lowered to meet the demands of war. Yet this was not solely about manpower levels; the authorities also sought to impose discipline and re-militarise hospitals, which sometimes were seen as disruptive impediments to the successful waging of the war. The German War Office failed in its attempts to synchronise military demands on time with patients' recoveries, as medical professionals resisted military authorities. Military and medical time clashed, with this case study serving as a useful example of broader civil–military disjunction across German society during the war.

Anna Grillini ("Fast Therapy and Fast Recovery") similarly analyses attempts to coordinate civilian and military timescales through an examination of the Italian neuropsychiatric service during the war. As in Germany, Italian military authorities required men to recover from maladies as rapidly as possible so to maintain manpower at the front. This, rather than the patients' well-being, was the top priority, and in turn invasive treatments were used in a bid to speed up recovery times. These took place at neuropsychiatric 'villages' – essentially asylums – where men endured a 'timeless reality' in which the only certainty was an inability to escape from the war. This was underscored by the echoes of artillery from the nearby frontline. Italian medical professionals initially viewed this novel context as an opportunity to push the boundaries of their field and explore new avenues of psychiatry. However, their hopes were dashed as the exigencies of war slashed the amount of time doctors could spend treating patients. In turn, they succumbed to pressure from military authorities, choosing the army and the national cause over medical professionalism, becoming increasingly militarised. They continued to focus on faradisation, including when differentiating between 'imposters' and 'legitimate' cases of psychological trauma.

Treatment was aimed at recovering soldiers rather than humans, and they worked to clear out their 'villages' swiftly, so as to prepare for the incoming wave of patients following the next battle.

Steve Marti's chapter ("A Stitch in Time") provides new insights into a hitherto overlooked aspect of the role of Britain's dominions in the war: material contributions from the home front. These contributions are juxtaposed with the better-known deployment of manpower from the dominions of Canada and Australia to Europe. Some of these voluntary efforts were substantial, including purchases of aircraft and equipping hospitals. However, much of the material contribution, which took the form of items such as clothing, was hampered by the distance between Europe and these dominions. The time it took to ship items created by social groups in the dominions meant that they failed to keep pace with the war, except in the most predictable cases, such as sending warm clothes for winter weather. While these contributions were therefore less effective than they might otherwise have been – with time wasted on the home front as a result – the groups which produced these items resisted both advice that they send money so that goods could be purchased in Europe, and attempts to centralise and streamline their efforts by national authorities. These issues were further exacerbated by the insistence of local groups that their contributions reach soldiers from their area. In turn, new questions are raised about the limits of imperial loyalties and the importance of local patriotism in the dominions during the First World War.

Cedric Van Dijck's chapter ("Slow Going") draws on renewed historiographical interest in print culture – especially the trench press – and complicates the relationship between these activities and the experience of time during the conflict. Focusing on the parallel experiences of Guillame Apollinaire and Virginia Woolf, his transnational analysis explains how boredom proved to be the key motivating factor in these pursuits, as well as a desire to use creative outlets to retain a sense of self during the dehumanising experience of war. Turning his lens to 'ugly feelings', Van Dijck explores private archival records to rethink the historical context of the emergence of modernism. For both Apollinaire at the front and Philip Woolf – Virginia Woolf's brother-in-law, who was wounded in combat – engagement with printing proved therapeutic, and a means for controlling the passage of time. This came in response to the stalemated warfare of the Western Front, which seemed to challenge contemporary notions of speed and improvement.

Part II switches focus to examine the relationships of time, most notably its relationships with experience and memory. Assaf Mond's chapter ("War-time, Zeppelins, and Children's Experience of the Great War in London") argues that the Zeppelin raids on London had such a significant impact on people's daily urban experience that their perceptions of time were altered. Most notably, people's responses to the Zeppelin threat blurred the boundaries between day and night, and Londoners changed the timeframes of their daily lives in response. Mond then builds upon this to demonstrate how these changes to time on a daily micro level had broader macro effects as well: disrupting the standard timeline of maturing into adulthood. The contours of the "distinct worlds of children and adults in Great War London" were erased. The author draws extensively upon sources created by children themselves to demonstrate that their accounts can enrich our understanding of the Zeppelins' effects upon time perceptions and the resulting altered experiences of Londoners.

Robert Rennie's chapter ("Time, Space, and Death") explores how the relationship between time and memory impacted identities and perceptions of German fighter pilots, during both the war and its aftermath. Three types of memory are explored that impacted these identities at different times: during the war, older tropes of medieval combat were resurrected and repurposed to make air fighting comprehensible; after death, individual memory of killed aviators created memorialisation in a manner distinct from that of trench deaths; and after the war, collective memory of both living and killed fliers were "re-purposed, re-buried, and rhetorically reconstructed, as either representations of an earlier time or standard bearers for a new, National Socialist Reich". Rennie also discusses how not just the memories of aviators but also aviators themselves were sharply influenced by time. Aviators experienced a regimented implementation of time during the war, interspersed with heightened moments of crisis when killing in the air, and this created a profound but brittle identity for Germany's fliers.

Erica Grossi's chapter ("The Photo Albums of the First World War") also examines the relationship between time and memory, but through the lens of material culture. Picking up on themes of pacing from Part I, by arguing that photography had the capacity to freeze time and suspend the 'dangerous moment', Grossi demonstrates how photograph albums served to display the materiality of time through images and to activate and reactivate memory of the conflict. With a particular focus on Italian case studies, the chapter takes an anthropological approach to positioning the photo album as a device used to familiarise the human gaze with the destructive character of this modern conflict and to practise the visual materiality of historical time. The chapter posits that in multiple 'presents' (collective, national, and individual), photo albums have served to activate a material, fragmented, and suspended past of the first 'war time' of the twentieth century.

Part III examines how contemporaries interacted with the temporal grammar of past, present, and future. Ashley Garber's chapter ("Brothers – and Sons – in Arms") discusses the memory of the Great War as construed by the members of the British and American Legions in relation to their Second World War counterparts. In this comparative analysis, Garber explores how ex-servicemen of the Great War renegotiated their past in order to come to terms with the present and with the new generation of veterans. Through an examination of periodicals of the British and American Legions, the author focuses on veterans' rhetoric surrounding the topics of war aims, service life, and kinship. She argues that both organisations sought to bring together different age groups by observing similarities in their respective war aims and service life, as well as by utilising the rhetoric of kinship. In turn, Garber makes a convincing argument for the incorporation of a relational, intergenerational focus into historical studies in order to facilitate a deeper and more nuanced understanding of questions of identity formation.

In his chapter entitled "Between Passatism and Futurism" Sante Lesti provides readers with a fascinating study of the war time and early post-war perception of time, as reflected in the rites of the Catholic Church. The author focuses on six main rites of consecration to the Sacred Heart of Jesus, which took place in different parts of Europe. Drawing on a number of sources in several different languages, Lesti discusses the rites in relation to three different orders of time – the Christian order of time, the Herodotean order of time, and the classical order of time. In turn, he argues that these rites served as a vehicle for re-establishing continuity that had been interrupted by the war. According to Lesti, they related the present, the past, and the future, and he goes on to argue that these rites depicted the conflict as a crucial turning point in this relationship. Furthermore, Lesti believes that the perception of the past and the future imbued in the rites made the present more understandable.

Finally, Alex Mayhew's chapter ("Hoping for Victorious Peace") explores the hitherto largely overlooked topic of hope and visions of peace amongst English soldiers in the Great War. In this chapter, the author discusses the importance of hope for a peaceful future for soldiers' morale and thus allows us to better understand the behaviour of men who found themselves surrounded by the horrors of the war. Although the perceptions of peace varied amongst the men, Mayhew traces the dominant similarities in their rhetoric. He argues that hopes for peace, in which the optimistic future was based on memories of the past, allowed the men to distance and abstract themselves from the terrible present. Moreover, the very acts of hoping and investing oneself in the future are recognised as being endowed with important psychological benefits. Furthermore, by believing that the armed forces were moving towards this goal, which was only possible through the defeat of despotic Germany, the soldiers found themselves imbued with a sense of purposefulness, thus reinvigorating their support of the war effort.

Despite the best attempts of its editors, the essays in this volume focus almost entirely on the Western world, with the result that a number of important scholarly areas of study remain unaddressed. War time and its perception, role, and position in non-Western cultures, religions, spaces, and combat experiences are only some of the fascinating yet regrettably unaddressed disciplines that come to mind. Additionally, many topics related to the Western world have not been exhausted and also remain open to investigation. As the diversity of essays included in this book suggests, time, its perception, and its roles can be examined in many different ways and from a wide variety of angles. The material history of time, time and operational history, and time and gender particularly stand out among the numerous avenues worth encouraging for further investigation. Nevertheless, all ten chapters in this volume explore the Great War from a novel and stimulating perspective. The volume allows emerging scholars to play a role in defining a new and crucial field in the study of the Great War. Yet if the conflict in question had once been deemed 'the war to end all wars', this volume has no similar aspirations with regard to further scholarly study of time. Its editors hope that the volume will inspire and stimulate further research in the field, which has been long overdue.

Notes

1 Bruce Scates, *Return to Gallipoli: Walking the Battlefields of the Great War* (Cambridge: Cambridge University Press, 2009); Tom Sear, "Uncanny Valleys and Anzac Avatars: Scaling a Postdigital Gallipoli", in Raelene Frances and Bruce Scates (eds), *Beyond Gallipoli: New Perspectives on Anzac* (Melbourne: Monash University, 2016), 55–81; Brad West and Haydn Aarons, "War memory, national attachment and generational identity in Australia", *Journal of Sociology* 52:3 (2016), 586–604.

2 Daniel Rosenberg, and Anthony Grafton, *Cartographies of Time* (New York, 2010).

3 Cheryl A. Wells, *Civil War Time: Temporality and Identity in America, 1861–1865* (Athens: University of Georgia Press, 2005).

4 Paul Fussell, *Wartime: Understanding and Behaviour in the Second World War* (Oxford: Oxford University Press, 1989).

5 Mary Dudziak, *War Time: An Idea, Its History, Its Consequences* (Oxford: Oxford University Press, 2012).

6 Lee Jarvis, "Times of Terror: Writing Temporality into the War on Terror", *Critical Studies on Terrorism* 1:2 (2008), 245–262.

7 See, for example, the chapter "Time and History" in Reinhart Koselleck, *The Practice of Conceptual History: Timing, History, Spacing Concepts*, trans. Todd Samuel Presner et al. (Stanford: Stanford University Press, 2002), 100–114. See also Reinhart Koselleck, *Futures Past: On the Semantics of Historical Time,* trans. Keith Tribe (Cambridge, MA: MIT Press, 1985).

8 François Hartog, *Regimes of Historicity: Presentism and Experiences of Time*, trans. Saskia Brown (New York, 2015).

9 Some notable new works on the United States include Michael S. Neiberg, *The Path to War: How the First World War Created Modern America* (Oxford: Oxford University Press, 2016); David Woodward, *The American Army and the First World War* (Cambridge: Cambridge University Press, 2014); Thomas W. Zeiler, David K. Ekbladh, and Benjamin C. Montoya (eds), *Beyond 1917: The United States and the Global Legacies of the Great War* (Oxford: Oxford University Press, 2017).

10 For two insightful historiographical reviews of recent studies see Heather Jones, "As the Centenary Approaches: The Regeneration of First World War Historiography", *The Historical Journal* 56:3 (2013), 857–878; Alan Kramer, "Recent Historiography of the First World War (Part I)", *Journal of Modern European History* 12:1 (2014), 5–27; Alan Kramer, "Recent Historiography of the First World War (Part II)", *Journal of Modern European History* 12:2 (2014), 155–174.

11 Stephen Kern, *The Culture of Time and Space: 1880–1918*, revised edition (Cambridge, MA: Harvard University Press, 2003).

12 Modris Eksteins, *The Great War and the Birth of the Modern Age* (London: Bantam, 1989); Paul Fussell, *The Great War and Modern Memory* (London:

Oxford University Press, 1975); Samuel Hynes, *A War Imagined: The First World War and English Culture* (London: The Bodley Head, 1990), particularly 423–439; Jay Winter, *Sites of Memory, Sites of Mourning: The Grear War in European Cultural History* (Cambridge: Cambridge University Press, [2005] 2014), particularly 2–5.

13 Robert Gerwarth, *The Vanquished: Why the First World War Failed to End, 1917–1923* (London, 2016). A similar point is made in Gerwarth's and Erez Manela's article "The Great War as a Global War", in which the authors call for the use of a framework in scholarly research that is both temporally and spatially larger than the one commonly used. See Robert Gerwarth, and Erez Manela, "The Great War as a Global War: Imperial Conflict and the Reconfiguration of World Order, 1911–1923", *Diplomatic History* 38:4 (2014), 786–800.

14 Katrin Bromber, Katharina Lange, Heike Liebau, and Anorthe Wetzel (eds), *The Long End of the First World War: Ruptures, Continuities, and Memories* (Chicago: Chicago University Press, 2018).

15 John Horne, "How Many Beginnings? Entering the Great War, 1914–1917", Leverhulme Visiting Professorship seminar paper at the University of Oxford's 'Globalising n Localising the Great War' seminar, 12 October 2017.

16 Dominik Geppert, William Mulligan, and Andreas Rose (eds), *The Wars Before the Great War: Conflict and International Politics Before the Outbreak of the First World War* (Cambridge: Cambridge University Press, 2015).

17 Salim Tamari addresses this fascinatingly in the section "Living in Three Time Zones: Ottoman/Gregorian/Hijri" of his book *Year of the Locust: A Soldier's Diary and the Erasure of Palestine's Ottoman Past* (Berkeley: University of California Press, 2011), 35–38. He discusses the disjointedness created by daily transitions between public and private methods of timekeeping that followed different time zones, including the role played army regulations in contributing to the situation.

18 "Blood Swept Lands and Seas of Red", Paul Cummins and Tom Piper, www.hrp.org.uk/tower-of-london/history-and-stories/tower-of-london-remembers/#gs.ogkM=3I;*Europeana14–18*, European Union, www.europeana.eu/portal/en/collections/world-war-I; "We're Here Because We're Here", Jeremy Deller, https://becausewearehere.co.uk/.

19 Joan Beaumont, "The Politics of Memory: Commemorating the Centenary of the First World War", *Australian Journal of Political Science* 50:3 (2015), 529–535; Catriona Pennell, "Popular History and Myth-Making: The Role and Responsibility of First World War Historians in the Centenary Commemorations, 2014–2018", *Historically Speaking* 13:5 (2012), 11–14; Hew Strachan, "First World War Anniversary: We Must Do More than Remember", *The Telegraph*, 11 January 2013: www.telegraph.co.uk/history/9795881/First-World-War-anniversary-we-must-do-more-than-remember.html.

20 Catriona Pennell, "Learning Lessons from War?" Inclusions and Exclusions in Teaching First World War History in English Secondary Schools", *History and Memory* 28:1 (2016), 36–70.

21 E.g. Paul Cornish, "Imperial War Museums and the Centenary of the First World War", *Twentieth Century British History* 27:4 (2016), 513–517; Rebecca Clare Dolgoy, "Re-Ordering Bedlam: Mobilising the Past at London's Imperial War Museum", forthcoming article; James Wallis, "Commemoration, Memory and the Process of Display: Negotiating the Imperial War Museum's First World War Exhibitions, 1964–2014", PhD thesis, University of Exeter 2015.

22 Brian Bond, *The Unquiet Western Front* (Cambridge: Cambridge University Press, 2002) 49.

23 Basil Liddell Hart, *Reputations: Ten Years After* (Boston: Little, Brown, and Company, 1928), 147; David Lloyd George, *War Memoirs Volume I* (London:

Odhams, 1938), 323, 891; Robin Prior, *Churchill's World Crisis as History* (London: Croom Helm, 1983), 215, 263, 281.

24 Alan Clark, *The Donkeys* (London: Hutchinson, 1961), 22, 43, 87; Joan Littlewood, *Oh! What a Lovely War* (London: Methuen, 1965)

25 John Terraine, *Douglas Haig: The Educated Soldiers* (London: Leo Cooper, 1963), x, 384.

26 Paul Fussell, *The Great War and Modern Memory* (Oxford: Oxford University Press, 1975).

27 William Philpott, *Bloody Victory: The Sacrifice on the Somme and the Making of the Twentieth Century* (London: Little, Brown, 2009); *War of Attrition* (New York: The Overlook Press, 2014); Gary Sheffield, *The Chief: Douglas Haig and the British Army* (London: Aurum Press, 2011); *Forgotten Victory: The First World War: Myths and Realities* (London: Headline Book Publishing, 2001).

28 Louis Halewood, "'A Matter of Opinion': British Attempts to Assess the Attrition of German Manpower, 1915–1917", *Intelligence and Nations Security* 32:3 (2017), 333–350.

29 A.J.P. Taylor, *War by Timetable: How the First World War Began* (New York: American Heritage, 1969), 28.

Part I
Speed, pacing, and suspension

1 No time to waste

How German military authorities attempted to speed up the recovery of soldiers in home-front hospitals, 1914–1918

Alina Enzensberger

In the first sentence of his essay "Advice to a Young Tradesman, Written by an Old One", Benjamin Franklin states: "Remember that time is money."[1] Equating time and money has since become a popular phrase. The final paragraph of Franklin's short essay is less well known. Here Franklin explained: "In short, the way to wealth [...] is as plain as the way to market. It depends chiefly on two words, industry and frugality; i.e. waste neither time nor money, but make the best use of both." Saving time (and thereby money) instead of wasting it – this catchy and yet trivial maxim has not only been predominant in modern industry, business, banking, and even private life, but also in waging modern war. It certainly played a major role in the First World War, when time and speed became crucial factors in military tactics. This was true for all belligerent armies, but particularly for Imperial Germany.

The German military had premised the so-called 'Schlieffen plan', its initial attack strategy, on the powerful effects of superior tactics and a rapid success by overwhelming the war enemies on the Western Front before heading to the East.[2] However, this manoeuvre famously failed, and soon the pace of the conflict on the Western Front changed as well.[3] By mid-November 1914, the war of movement had come to a halt. It gave way to an exhausting trench warfare producing high costs on all sides, both in human lives and in material resources.[4] This was especially true for the battles of Verdun and the Somme in 1916 when both sides conducted attritional campaigns to wear down the enemy's resources – without eventually being able to decide the war.[5] While being locked into this stalemate was a devastating situation for all warring parties, it was particularly consumptive for Imperial Germany. As the German troops were clearly outnumbered by the Allied forces from the very beginning of the conflict until 1918, it became obvious that in the long run, regardless of superior military equipment and innovative capacity, the mere number of soldiers fit for action and the workforces in the industry would eventually be the decisive factor.[6] Additionally, due to the Allied blockade and to an excessively bureaucratic, inefficient administration in Germany, material conditions at the home front deteriorated drastically after 1916.[7] As a result, prices rose, the quality of goods deteriorated, and replacement products were increasingly needed to fill the gaps. These shortages led to disastrous consequences for the health condition of the population, especially in larger cities, and thus to an aggravated crisis of morale at the home front. Yet fading domestic support was the last thing military leaders needed, as this support remained essential for continuing the war. Something needed to happen, as time was playing against the German Empire.

This increasingly pressurised situation of a ticking clock established its own competitive logics, priorities, and timeframes. On a national level, German authorities reacted with enhanced mobilisation measures such as the Hindenburg programme in autumn 1916, the Auxiliary Service Act in December 1916, and unrestricted submarine warfare in February 1917, as well as intensified propaganda.[8] But how did the time regimes of trench warfare affect other areas of society? How did they change work routines and institutional functions? How did they influence facilities operating between military and civil life? In approaching these questions, this essay concentrates on military hospitals at the home front – an area of society that was located far away from the war events at the Western Front, yet inextricably linked to them.[9] How did the perceived time pressure of the army affect hospitals at home?

It has frequently been argued that the First World War not only unleashed a formerly unknown dimension of mass killing since the first weeks of combat, but also produced unprecedented forms of physical injuries, illnesses, and psychiatric disorders in all belligerent armies.[10] Weapon technology had developed rapidly, especially with the experiences of the battles of Verdun and the Somme in 1916. At the same time, medical technology became increasingly modern, quick, and much more effective. Yet the soldiers' bodies did not 'modernise'. This led to a temporal discrepancy between modern wounding and human healing: it was a matter of seconds to inflict severe wounds on a soldier's body, but it would take the injured man weeks, months, or sometimes years to recover in military hospitals. Over the course of the war, temporary hospitalisation became a mass phenomenon, concerning millions of servicemen.[11] At the same time, the physical condition of soldiers was increasingly considered to be one of the decisive factors in eventually defeating or at least holding out longer than the enemies.[12]

Under these precarious circumstances, German military officials made many attempts to conceive efficient medical treatment schemes in order to send patients back to the front or to work in the war industry as soon as possible. Despite their efforts, in the view of the Prussian War Ministry and other medical offices, convalescence time still took too long – particularly in hospitals at the home front. Since around spring 1915, officials were increasingly unwilling to accept long hospital stays.[13] In their view, the pressing requirements of the front and the war industry made it necessary to discharge patients considerably earlier. How did the military attempt to enhance its competitive time regime in hospitals at home? How did this influence treatment objectives and perceptions of health?

In order to answer these questions, I will concentrate on the perspective of military medical authorities and their internal and official documents.[14] I will focus on two major decrees, one from 1916 and the other from 1917. Both were purposefully issued to speed up patients' discharge. An analysis of these two cases reveals explicit and implicit objectives of military medical officials, which made them so obsessively insist on short-term hospital stays. I will argue that the constant claim for acceleration of patients' discharge entailed more than just the demand for more troops at the front and manpower in the industry. Firstly, it was not only based on economic and military considerations, but equally importantly also served as a rhetoric to discipline and re-militarise hospital surgeons and patients alike. By describing treatment time as a moral issue determining victory or defeat in war, authorities attempted to put doctors in their place, defining the hospital as a mere instrument of the military without other, that is, civil self-interests. Secondly, acceleration of patients' discharge, which became the most prominent claim of medical authorities during the whole course of the war, also seems to have served as a default reaction helping to give a simple answer to various complex problems in the hospital system at home and thus demonstrating administrative sovereignty, resoluteness, and activity.

Acceleration and synchronicity: reframing medical time

Conceptualising 'acceleration' in sociological terms, Hartmut Rosa offers a helpful perspective on these cases. According to Rosa, social acceleration is a fundamental aspect of modernisation: "The question, who decides over the rhythm, duration, pace, sequencing and synchronising of events and activities, constitutes a major field for conflicts of interest and struggles for power."[15] Rosa directs our attention to the problem of synchronisation. This occurs when processes of different velocity are to be coordinated in order to function together: "As soon as two processes intermesh, that is synchronise, the quicker element puts the other under time pressure – if the slower element does not accelerate accordingly, it is perceived as an annoying impediment, an operational limitation."[16] This general observation holds true for the context of hospitals in the First World War as well. Although the military authorities acknowledged that hospitals at the home front were essential rehabilitation instruments for soldiers, they also perceived them as disruptive impediments. Their attempts to synchronise a military time regime with a clinical and a patients', that is physiological, one failed – the seemingly more efficient military way of solving problems could not fully be carried through. Thus, the authorities started to increase the pressure. How did they do that?

Military officials considered a maximum of two months to be an adequate and unsuspicious time span for recoveries in hospitals.[17] The fact that healing from severe war wounds or illnesses would often take considerably longer than this was something they seemed to be reluctant to accept and even less willing to finance. In their correspondence and decrees, officials conceptualised hospitals as reservoirs of future servicemen, not yet exploited to their full potential but impatiently waiting to be made available for the army again. By focusing on the procedure of discharge from hospital, they framed their speed-up demand as an issue of process optimisation, and thus insinuated that extended hospital stays were mere procedural errors. As such, they could be named, clarified, and then erased.

By repeatedly addressing the issue of overstay, authorities also conveyed – and this was primarily a message to their medical staff on site – what the main task of hospitals was and should be: it was not a medical one in the civil sense of the term, aiming at healing the soldier and restoring his health as fully as possible, but a military medical one. Hence, patients were meant to be treated on a trial basis. Hospitals were supposed to investigate how fit the injured would become in a reasonable amount of time. If this process took disproportionately long, they were to be discharged. In this case, civilian agencies had to take charge of their subsequent rehabilitation. Additionally, officials made many efforts to integrate them back into the labour market at the end of their hospitalisation.[18] Before being discharged from hospital, every soldier was assigned one of four fitness categories in order to classify his bodily capability: 'k.v.' (*kriegsverwendungsfähig*) meant fit for active service, 'g.v.' (*garnisonsverwendungsfähig*) fit for base duty, 'a.v.' (*arbeitsverwendungsfähig*) fit for work duty and finally 'd.u.'/'kr.u.' (*dienstunbrauchbar/kriegsunbrauchbar*) unsuitable for service.[19] These four categories defined where the soldier could be used next. In other words, the main mission of military hospitals was not to perfectly cure the patients' bodies in a clinical sense, but to try to restore either their military fitness or their economic capability, at least what was left of it.[20]

Yet, many physicians committed to hospitals at home had been civilian doctors in their pre-war lives and were not used to this way of thinking.[21] Although research has shown that most hospital doctors were resolved to support their country's war effort and although they had the rank of medical officers, they did not all automatically acquire the self-conception of military surgeons and change all their pre-war priorities and Hippocratic work morale.[22] This role conflict of many military surgeons and consequently of hospitals as their places of action soon manifested itself in the acceleration debate discussed here.[23]

Hospitals as problematic areas: why long treatment periods created concern

Who did the military authorities hold responsible for causing the deceleration of soldiers' discharge from hospital? And more generally speaking: why were they so desperate to get rid of patients as soon as they possibly could – including even men who would clearly not return to the war front? Let us start with this second question, which makes relevant another question: why were extended hospital stays seen as a major problem in the first place?

When reading through internal and official documents of the War Ministries, subordinated medical offices, or even scientific papers by surgeons, four main points of criticism stand out. Firstly, as previously mentioned, the army was in a constant struggle to recruit enough soldiers fit for some form of military service – either at the front itself or in a position behind the lines that did not require high levels of fitness. From a strategic point of view, it appeared virtually self-destructive to stand idly by watching precious fighting capacity seeping away into hospital care for an indefinite time. This became ever more urgent during the course of the war. In August 1916, a medical superintendent from Baden stressed the urgency to quickly send men back to front service by demanding that injured soldiers not be handled with kid gloves:

> I consider it necessary that [...] in this current state of a high demand for servicemen the standards should not be raised too high and that according to all experience even men with chronic bronchitis, with little excretion of protein without cylindruria, with mitral incompetence but otherwise good general condition are by all means up to trench warfare and must not be sent homewards.[24]

Other officers took this way of arguing even further and implied that field service was not a risk after hospitalisation but actually "good for a soldier's further convalescence [*genesungsfördernd*]".[25]

The second point of criticism was again a matter of military organisation and directly related to the first. Authorities at the home front were concerned about losing control over patient distribution when hospitals were working to capacity. More precisely, they feared being confronted with a situation of ambulance transports arriving unexpectedly and not finding enough free bedsteads right away, because all hospitals were overcrowded with old cases and not flexible enough to react to new requirements. This concern was not completely unfounded. Similar problems had indeed arisen in the chaotic beginning of the war when hospitals at home were not yet fully established and not experienced in handling big amounts of new patients.[26] Certainly, losing control again was not something authorities wanted.

Thirdly, in the strained financial situation Germany was facing, authorities generally sought to avoid costs by all means. This included the medical realm as well. Hospital stays were financed by the military administration. In this way, soldiers did not have to pay treatment in military hospitals themselves.[27] As soon as patients were discharged from the clinic and dismissed from the army as unfit for service, the medical fund was not accountable for them anymore.[28] However, even during hospitalisation, authorities wanted to save money by saving time. On the one hand, they sought to keep down direct costs such as those for medical material, catering, and staff, as well as for renting buildings. On the other hand, they also tried to sidestep indirect expenses. The idea behind this was the following: if an injured man was predictably not fit for field service anymore but able to work in the industry, he was still regarded as useful to the war effort. In this case, the invalid could take over the job of a present industrial or agrarian worker fit for duty. Inversely, that former worker or farmer could be conscripted to the front and would fill the gap the patient had left.[29] However, if hospitals stood in the way of this procedure of role swapping by clinging to their patients, this would in the long run seriously weaken the war industry. According to Dr Karl von Seydel, chief of the Medical Department (*Medizinal-Abteilung*) in the Bavarian War Ministry, it was indeed "an urgent interest of our fatherland not to allow any potential workforce, how small as it may be, to lie fallow longer than strictly necessary. Ways must be found to deal with this issue of upmost importance."[30]

Last but not least, authorities feared that long-term hospitalisation at the home front would distance patients from military life and cause social unrest. This was a different type of concern, albeit possibly the strongest of all, since it was constantly repeated in many variations. Authorities feared losing control over patients' combat readiness and general state of mind if they allowed them to stay in reserve hospitals for too long. As much as they considered these clinics to be indispensable military tools, they also viewed them as highly problematic, somewhat inscrutable places. Therefore, hospital stays always seemed to carry the risk of causing collateral social damage.[31] Dr Karl Frickhinger, a councillor and medical officer in Würzburg, urgently warned against extended hospitalisation from a psychological point of view:

> A long-term hospital stay caused by severe injuries and illnesses bears its own risks of which the worst is the long-lasting inactivity. The sick person is weaned from work, his thoughts revolve way too much around his physical condition and well-being. As a result, the whole consciousness egocentrically concentrates more and more on the beloved self. If, on top of that, the patient is coddled and pampered by rather inexperienced nursing staff, then we make the same experiences in military hospitals that we find so often in home care. For the psychologically not completely unimpaired we slowly lay the ground for a psychoneurosis.[32]

Statements like this reveal that medical officers did not only view the quick discharge of patients as an abstract economic necessity. At the same time, the speed-up demand could be used as a powerful administrative tool in order to get rid of neurotic or recurrently sick patients – which were often considered as the same. In this way, medical officers hoped to counter the danger of waning discipline in hospitals at home. Willy Hellpach, a military surgeon and later politician of the left-liberal party DDP, postulated in a journal article in 1915 that patients should by no means be pampered in hospital as they risked being weaned from their military and professional duties:

> If we radiate too much convenience, too much cosiness, too much comfort, too much distraction in hospital, then this medical institution eventually becomes a club or society house, where the purpose of [the

> soldier's] treatment, the recreation of his fitness for duty, vanishes. It's true, the soldier should have the feeling, that it's nowhere as pleasant as afield, despite all danger and exertions, and nowhere as little pleasant as in the military hospital, despite all security and rest.[33]

For many soldiers, however, Hellpach's vision had little to do with their own hospitalisation experience at the home front. Actually, quite the opposite was true in most cases. Ironically, in a situation of technologised trench warfare, disease and wounds could mean survival, whereas physical health potentially put a soldier's life in great danger. While field hospitals close to the front line were mostly provisional, chaotic, and sometimes highly dangerous places where the horrors of the war became blatantly obvious,[34] hospitals back home could be very comfortable and welcoming. As numerous patients' diaries and letters show, soldiers had – often for the first time in months – their own clean bed with white linen; they found rest, a calm environment, enough food (except for periods of extreme food shortages), and they came into more regular contact with women, above all with nurses. Moreover, visitors were allowed in hospital. This included family members and friends as well as external visitors from the city in which the hospital was situated. Additionally, patients with only a benign condition or who were on the mend often had many possibilities to move freely and to organise their leisure time. Diaries of these types of patients reveal how many different everyday activities were offered by hospital personnel or initiated by soldiers themselves in order to counter boredom and melancholy. Common activities were trips to nearby cities, going to the cinema or to the theatre, playing theatre in a patient drama group, and playing card and board games.[35]

It is not surprising that military authorities kept a wary eye on these forms of leisure activity happening within their own medical institutions. Especially from 1916, with the radicalisation of the war on the Western Front, they refused to accept any additional time loss from soldiers assumedly lingering in hospitals. Dr Philipp Hofbauer, head of the Medical Bureau [*Sanitätsamt*] I. Army Corps in Munich, insisted on this point in a letter directed to all reserve hospitals in his area of competence: "A long-term stay in hospital brings about nothing but adverse effects; dread of service and pension addiction are raised, general indolence fostered. As a result, the much needed workforce in the army, in the industry and in agriculture is lost."[36]

Altogether, the four main problems which authorities saw arising from delayed discharge were: firstly, manpower shortage at the front and in the industry; secondly, lack of space in hospitals; thirdly, superfluous costs; and fourthly, loss of military discipline and labour resistance of patients with all the related psychological and social problems. While medical officials seemed to agree on all of these points, it was more complex for them to grasp the reasons for the observed length of treatment. Indeed, many internal letters only addressed this issue and admonished hospital physicians to speed up the discharge of patients, without offering concrete instructions how to do so.

In search of the culprit: the first acceleration decree of 1916

In July 1916, the Medical Department of the Prussian War Ministry attempted to remedy this grievance once and for all. To this end, it organised a major assembly of all war medical inspectors (*Kriegs-Sanitäts-Inspekteure*), deputy corps physicians, and other representatives of medical offices in the 'Kaiser Wilhelm Academy for the education of military surgeons' and put up a new decree for discussion. One month later, Surgeon General Wilhelm Schultzen, head of the Prussian Medical Department, finally released the seminal enactment with the long title "Considerations about the question of an accelerated conscription of men fit for active service or fatigue duty from military hospitals and units".[37] It was classified as 'for official use only'. The decree consisted of three main parts, each subdivided into several "reasons for delay" and corresponding "remedial measures". In total, it listed no less than twelve overarching reasons for delay, often comprising several additional obstructing causes on a more detailed level. This particular design of the decree is quite instructive for the reader of today as it focuses – probably more than the Prussian Medical Department intended – on the dysfunctional aspects of hospital treatment and medical examinations at the home front. It helps to understand which points of criticism high-ranking military surgeons deemed important or communicable and what was left out. In this way, the decree offers an interesting insight to a panorama of internal conflicts and misunderstandings between military medical authorities and hospital doctors. Moreover, it was a document of bureaucratic self-reassurance, which insinuated that procedures could be controlled and accelerated if they were only commanded in a clear and resolute manner.

The reasons for delay addressed by the new acceleration decree can be categorised in three main groups. The first type shifted the blame to hospital physicians and nursing staff by addressing assumed procedural violations. This included, for example, improper list keeping, incomplete or incorrect patient documentation, and equivocal medical reports. As the decree stated, it was this lack of administrative accuracy in hospitals that made it necessary to constantly recheck patients' information by contacting other offices and hospitals. As a result, valuable time was wasted since men fit for service were stuck at the home front. In other cases, the decree went on, soldiers were sent back and forth between the hospital and the army physician since both doctors refused to trust and follow the medical evaluation of the other. These, the decree concluded, were unnecessary homemade problems that needed to stop immediately.

The second type of reasons for delay was intimately linked to the first and concerned procrastination due to uneasiness with military bureaucracy and its formal demands. According to the Medical Department, doctors often delayed issuing medical certifications because they were not used to, or comfortable with, formal military reporting:

> Issuing certifications is often not the strength of civil doctors with their particular professional background; many times the military medical

> evaluation of individual cases is complicated [...]; formalities are emphasised more than necessary by certain offices, rejection of certificates for formal reasons increases the reserve and reluctance towards this task, encourages mutual buck-passing of issuing certifications.[38]

At this point, the Medical Department also included some self-critical sentences: some offices, it admitted, might have expected too much bureaucratic perfection of doctors who were not trained for this, although it was inadequate to meticulously insist on unblemished certificates in a situation of significant time-pressure. Therefore, the decree ordered, medical bureaus should accept every single certificate without further fuss if it offered "sufficient" and "convincing" information about a patient's identity and reason for discharge from hospital.[39]

The third and final reason for delay was the alleged personal bias of doctors. According to the decree, some physicians with a lax work ethic, especially in small auxiliary hospitals in the countryside, kept patients indefinitely in their ward with vague diagnoses and without any clear therapeutic goals. This situation even worsened if the patient happened to be lying in a hospital of his hometown: "In hometown hospitals unfavourable influence of relatives upon the sick in terms of delay of rehabilitation; here again breeding of neuroses",[40] the decree stated concisely. Moreover, it not only accused some doctors of paying too much attention to certain wishes of the patient's relatives but also blamed others of pursuing their own self-interests instead of the nation's: "Decisive for longer retention of patients are often considerations for the military hospital itself, whose existence – in the case of small auxiliary hospitals – depends on the occupancy rate with patients."[41] Although the decree did not explain this observation in more detail, it actually constituted a big concern of the Medical Departments. Auxiliary hospitals – in German "*Vereinslazarette*", meaning hospitals run and sometimes financed by civil associations such as the Red Cross or private providers[42] – were under general suspicion throughout the whole course of the war.[43] They seemed to embody all the negative aspects of hospital care in a concentrated way, such as loss of discipline and combat morale among patients, effeminacy, misguided and unrealistic treatment objectives, and lack of military control. Concerning the delayed discharge of patients, this general mistrust now culminated in accusing auxiliary hospitals and their staff of clinging to their patients on purpose in order to pursue their own scientific and operating interests instead of unconditionally supporting the national acceleration effort. This, the decree insisted, was not only counterproductive, but seriously harming: "Physicians have to be reminded again and again that any retention of a patient in hospital that is not absolutely necessary weakens the combat power on the military battle field as well as on the economic one."[44] With this claim, the Medical Department deliberately framed the need for acceleration of hospital discharge as a question of national faithfulness and combat readiness. Inversely, if this message was thought through to the end, it implied that doctors, who actively or passively impeded the speed-up effort, were traitors to their fatherland. Here, at the latest, the argumentation had reached a moralistic level.

Beyond these three main arguments, the enactment named many more subordinate reasons for delay; there seemed to be endless variations of bureaucratic flaws, communication problems, and other sources of error, which all inhibited the acceleration success the military so desperately aimed for. What was only briefly mentioned here but played a major role in other documents was the role of nurses and patients in intentionally delaying the discharge process. Nurses and other women providing care such as relatives or fiancées were accused of pitying and pampering patients to an extent that they jeopardised their healing will and inner strength.[45] Patients, on the other hand, were often suspected of aggravating or simulating disease in order to delay front service, particularly, if the clinical picture was not clear.[46] Hence, more generally speaking, the Medical Departments did not only suspect uncooperative hospital doctors and obstinate offices to be potential time-eaters, but also caregivers and patients themselves. In fact, the group of presumed culprits encompassed almost all non-military actors involved in hospital service. However, the Medical Department did not address their share of guilt in greater detail in the acceleration enactment of 1916. It can be assumed that this was because the decree was not aimed at non-military actors as a readership. Instead, it was a warning message to the medical colleagues in hospitals, at least to those who had not yet fully adopted a military mind-set but still followed the principles of civil medicine with its corresponding slower time regime and different medical goal.

What the list of decelerations did not touch upon at all was the initial reason for delay, the elephant in the room: the nature of modern war itself. New types of weapons, especially artillery guns, were able to produce horrifying wounds, untypical in previous wars, such as facial destruction or severe soft tissue injuries, which took a long time to heal.[47] Moreover, the devastating conditions in the trenches also caused grave damage, such as skin diseases, frostbite, or kidney inflammations.[48] Being a soldier had become more dangerous than ever before.[49] At the same time, chances to survive had increased in other ways, as not only weapon technology had modernised but along with it the treatment methods of military medicine. Indeed, it was now possible to cure wounds that would have been fatal in former times, especially thanks to the discovery of aseptic wound care, X-ray units, and blood transfusion,[50] as well as mass-scale tetanus vaccinations, which altogether dramatically decreased the number of infections and casualties during operations. Nevertheless, in this progress competition between weapon technology and medical technology, both of which battled for the life of soldiers, medicine inevitably had to fall behind in the long run. Moreover, the price to pay for some of its spectacular healing results was long-term hospitalisation of wounded soldiers. In the German army, combat wounds were the most common reason for soldiers to be admitted to a military hospital.[51] Hence, what really made hospital stays so long and complicated in many cases were the severe wounds and illnesses caused by the war itself. But this, I argue, was something that could neither be changed quickly nor communicated in a positive manner. Therefore, both the decree discussed above and later enactments concentrated on the operational aspects of hospital treatment and discharge. Here, at least, some level of control could be reached.

What were doctors supposed to do to speed up hospitalisation? The Medical Department ordered numerous "remedial measures" in the decree, some of which have already briefly been mentioned above. They stood for several complementary strategies. The first was enhanced theoretical and practical instruction of hospital doctors by more experienced military surgeons. In this way, hospitals were supposed to increasingly fulfil the role of military training sites for doctors in order to bring them back into line. Secondly, the decree imposed measures in order to change the command structure in hospital medicine. They were supposed to both weaken the position of subordinate ward physicians towards their chief, and simultaneously strengthen the influence of external military actors such as war medical inspectors and medical commissions. From now on, these external assessors were supposed to survey hospitals on a more regular basis than before. During these visits, they were on the one hand expected to act as mentors and teach hospital doctors about military procedures. On the other hand, the decree empowered them to discharge patients at their own discretion, if necessary. Thirdly, the decree demanded to establish a stricter disciplinary regime in hospitals in order to foster the "recovery will" of patients. This was to be facilitated mainly by urging patients to work in hospital workshops or nearby factories instead of leaving them to themselves without surveillance.[52] Finally, recalcitrant auxiliary hospitals which had gravely impeded acceleration in the past were not to be tolerated any more. Regarding this issue, the decree did not offer any more adjusting "remedial measures" but called for radical action. Small, rural hospitals that opposed military rules should be closed without substitution, even at the risk of increasing the lack of beds at the home front.[53] There was no time for this kind of 'nuisance' any more.

Some of the measures cited above were not completely new. They had previously been ordered in letters and smaller decrees, especially concerning hospital workshops and instruction of doctors. What was innovative was the fact that they were for the first time in the course of the war assembled in a comprehensive decree in the form of a brochure with national reach, issued by the Medical Department of the Prussian War Ministry. Clearly, in July 1916, the demand for acceleration of hospital discharge had reached a level of national importance. Yet it seems that the decree was not able to fully reach all of its ambitious aims. As internal documents show, the authorities' complaints about excessively long hospital stays at the home front did not diminish after July 1916.[54]

More of the same: the second acceleration decree of 1917

Nine months later, on 12 April 1917, the Prussian War Ministry started its second major speed-up initiative. It issued a renewed acceleration decree, albeit this time not only signed by the Medical Department, but by war minister Hermann von Stein himself. As its title "Order for the accelerated provision of conscripts and members of the army fit for active service or fatigue duty by the district headquarters, units and military hospitals" was rather lengthy again, the War Ministry additionally suggested the abbreviated title "Discharge-acceleration-order" (*Entlassungs-Beschleunigungs-Anweisung*) or even more briefly "Eba".[55] In contrast to the decree from 1916, the "Eba" did not discuss reasons for delay any more – possibly because this had made the system's weaknesses too clear and thereby vulnerable to attack. Instead, it straightforwardly defined acceleration measures and timesaving goals. In terms of content, however, it did not differ much from the last decree. The War Ministry did not abolish any of the past measurements, but mostly repeated them in a harsher tone. Its main objective was already formulated in the second paragraph: "The quickest way for reaching a reliable medical decision is the best. Any formal concerns must stand back."[56] Most interesting about the "Eba" is not so much its content, but the way it was supposed to be taken note of by its recipients, especially in the case of Bavaria.[57]

When the Deputy General Command I. Bavarian army corps handed over the "Eba" to all subordinated units, it added a set of additional comments to some of the decree's paragraphs. Most importantly, it ordered two enforcement measures in its letter: firstly, the "Eba" was only to be handed out to doctors and military units against signature. Also, a list of all medical recipients had to be reported back to the Deputy General Command in order for it to gain an overview. Secondly, the garrison and chief surgeons were repeatedly supposed to gather all hospital doctors within their field of competence, read out the complete 45 pages of the "Eba" in front of them, and eventually add "useful tips regarding their previous service experience".[58] The letter insisted on this point: "A mere distribution is not enough, but it has to be ensured that the content of the Eba becomes the intellectual property of all doctors on duty."[59] In the future, the letter went on, the chief surgeons should gather the other doctors for joint conferences on a monthly basis. At these occasions, the instructions and experiences with the "Eba" were to be discussed further. It can be assumed that with methods like these, the military authorities, at least in Bavaria, wanted to make sure that the new acceleration decree would not fail to operate once again. It needed to make a stronger impression on doctors. Therefore, the chief physicians were advised to implement the "Eba" in this particular performative manner – instead of just handing it out mechanically – in order to ensure that hospital doctors would truly internalise it this time.[60] With this propaganda technique, authorities insisted more strongly on acceleration than ever before.

Until the end of the war, many of the remedial measures ordered by the "Eba" seem to have been carried through indeed. But nevertheless, in the view of military authorities, the situation continued to remain unsatisfying. In official letters of the Medical Departments and other military agencies from late 1917 to 1918, the complaints about long-term hospital stays and delayed discharge remained a major topic.[61] Towards the very end of the war, with the gradual disintegration of the German army, this issue became even more relevant. More and more field hospitals had to be removed so that an increasingly high number of injured and sick soldiers sought treatment in hospitals at the home front.[62] Altogether, it seems that even in a situation of 'total war', total acceleration remained an unreachable dream of military administration.[63]

Conclusion

This chapter has examined how the increasing pressure of military time and attritional warfare influenced other areas of German society. It has concentrated on military hospitals at the home front and investigated why military authorities insisted, almost obsessively, on shortening hospital stays at home and discharging soldiers as soon as possible. The analysis has shown that military and monetary considerations were the main reasons for this. However, I argue that the never-ending claim for acceleration also fulfilled several other, at least equally important, functions. Firstly, the claim for acceleration reflected the military authorities' attempt to display bureaucratic activism and resoluteness in a situation of fading military options. In this sense, it had a self-reassuring function. Secondly, it served as a rhetoric, which could be used as a universal weapon against all sorts of perceived nuisances. Speeding up the procedures of treatment and discharge seemed to be a reliable way to get rid of long-term patients or even of unmanageable auxiliary hospitals. Thirdly, the claim for acceleration had a mobilising and disciplinary function. The two decrees of 1916 and 1917 were meant to both (re-)militarise hospital doctors and, inversely, discipline and control patients. Thus, in home-front hospitals, time not only equalled money, but temporal considerations also involved other issues such as fighting morale and national interest. As one result of this, perceptions of health and fitness gradually shifted. In times of war, the Prussian War Ministry stated in September 1915, one could not expect the same quality of conscripts as in times of peace.[64] Following this maxim, German doctors were increasingly advised to lower their high standards when assessing the bodily constitution of convalescents and new recruits.[65]

During the whole course of the war, but especially since 1916, the military authorities tried to synchronise their own time regime with the medical one and to impose their competitive standards. However, they often faced unexpected resistance by doctors, nurses, and patients alike. Despite all disciplinary efforts, home-front hospitals seem to have developed their own dynamics which could not be fully controlled by military demands and regulations. According to Hartmut Rosa, infirmaries, as well as factories, prisons, or schools, can be considered the primal institutions of modern society where new time regimes are trained, practised, and eventually internalised – or rejected: "Following a strict time regime plays a major role in all of these institutions […] and appears to be one of their crucial disciplinary goals."[66] The German hospital at the home front, where military and civil medical logics with their different time schemes met and often clashed, is a good example of this. It is worth examining this particular social space in order to understand the consequences of conflicting temporalities between the war front and home front on the one hand, and military and civil medicine on the other. Finally, this can also inspire a reconsideration of the debate on 'total war' from a new, temporal perspective.

Notes

1 Franklin, "Advice to a Young Tradesman", 304.
2 Ehlert et al., (eds), *Der Schlieffenplan.*; Groß, *Mythos und Wirklichkeit*, 61–144; Mombauer, "German War Plans".
3 Herwig, *The Marne 1914*; Münkler, *Der Große Krieg*, 158–213.
4 Epkenhans, *Der Erste Weltkrieg*, 50–57, 83–86; Leonhard, *Die Büchse der Pandora*, 146–154, 157–183; Overmans, "Kriegsverluste"; Ziemann, *Gewalt im Ersten Weltkrieg*, 24–62.
5 Foley, "What's in a Name?"
6 Afflerbach, *Falkenhayn*, 198; Stevenson, *1914–1918*, 44–73.
7 Kramer, "Blockade and Economic Warfare"; Leonhard, *Die Büchse der Pandora*, 518–520.
8 Feldman, *Army Industry and Labor in Germany*, 149–196; on propaganda see Schmidt, *Belehrung, Propaganda, Vertrauensarbeit*; on submarine warfare Rojek, *Versunkene Hoffnungen*, 173–204.
9 The concept of German military hospitals as transitory spaces between war front and home front is discussed in greater detail in Alina Enzensberger, *Deutsche Lazarette an der Heimatfront*, unpublished PhD dissertation.
10 Hofer and Prüll, *War, Trauma and Medicine in Germany and Central Europe*; Lerner, *Hysterical Men*; Ulrich, "Krieg der Nerven, Krieg des Willens".
11 For contemporary statistics from the 1930s concerning patient numbers see Heeres-Sanitätsinspektion des Reichskriegsministeriums, vol. 3. Although this is the only comprehensive statistical record available concerning the German army, its figures are somewhat misleading, as the 'Sanitätsbericht' counted every commission to hospital as a separate 'case', even if the same men came to hospital twice. As a result, patients' numbers were distorted upwards.
12 Harrison, "The Medicalization of War – the Militarization of Medicine".
13 For some of these early acceleration claims see, for example, University Archive Heidelberg H-III-600/1: decree by the Medical Bureau [*Sanitätsamt*] XIV. Army Corps, 01.02.1915; Bundesarchiv – Militärarchiv (BA-MA) RM 120/129: decree by the Prussian War Ministry, no. 4601/3.15. C 1, 11.05.1915; Bayerisches Haupstaatsarchiv – Abteilung IV. Kriegsarchiv (BayKrA) Stellvertretendes Generalkommando I. Armeekorps Sanitätsamt/176: decree no. 47800 by the Medical Bureau I. Army Corps, 14.10.1915.
14 All military hospitals at the German home front were organised and commanded by the army medical service at home (*Heimatsanitätswesen*) which worked separately from the field medical service (*Feldsanitätswesen*), even if they were constantly interacting. The central authorities for the army medical service at home were the four Medical Departments (*Medizinal-Abteilungen*) of the War Ministries of Prussia, Bavaria, Saxony, and Württemberg, each led by a Surgeon General (*Generalarzt*). While these four departments operated independently in theory, the Medical Department of the Prussian War Ministry issued central guidelines on a regular basis which were generally followed by the others. On a regional level, the military hospitals of each Army Corps were administered and controlled by a

Medical Bureau (*Sanitätsamt*) of the respective Deputy General Command (*Stellvertretendes Generalkommando*). At the head of each Medical Bureau was a Deputy Corps Military Surgeon (*Stellvertretender Korpsarzt*). Finally, on a local level, every reserve hospital (*Reservelazarett*) was managed by a chief surgeon with military medical training or by a mixed commission (*Reservelazarettkommission*). The chief surgeon was also in charge of the various auxiliary hospitals (*Vereinslazarette*) affiliated to his reserve hospital. For more details on the complex system of military medical care within the borders of the German empire see Heeres-Sanitätsinspektion des Reichskriegsministeriums, *Sanitätsbericht über das Deutsche Heer*, vol. 1, 157–169; K.S.O., *27.01.1907*, 79–89.

15 Rosa, *Beschleunigung. Die Veränderung der Zeitstrukturen in der Moderne*, 36.

16 Rosa, *Beschleunigung und Entfremdung*, 99.

17 For an official statement on this see BA-MA PHD 6/197: Gesichtspunkte zur Frage der beschleunigten Herausziehung militärisch oder in der Kriegswirtschaft verwendbarer Mannschaften aus den Lazaretten und Truppenteilen, 28.07.1916, 6.

18 Kienitz, *Beschädigte Helden*, 193–204; Perry, *Recycling the Disabled*, 87–105.

19 Martineck, "Die Begriffe Dienstbrauchbarkeit (Kriegsbrauchbarkeit)."

20 See for example BayKrA MKr 18389: Bavarian War Ministry, answers to the Bavarian Landtag, no. 123; Schultzen, "Organisation der zahnärztlichen Hilfe im Kriege", 246.

21 According to the statistical data of the Sanitätsbericht, on the random reporting date of 15 July 1916, 2,097 of the 8,601 doctors employed in military hospitals within the German Empire were indentured civil doctors (these figures did not include doctors working in facilities of the Red Cross or similar). See Heeres-Sanitätsinspektion des Reichskriegsministeriums, *Sanitätsbericht über das Deutsche Heer*, vol. 1, 36; after the Auxiliary Service Act from 5 December 1916, these numbers increased further – see ibid., 37–38.

22 Prüll and Rauh, "Krieg und medikale Kultur in Deutschland", 7–29.

23 Bergen, "For soldier and state", 291–308.

24 Generallandesarchiv Karlsruhe 456 F 113 Nr. 82: Report by the chief physician of the military hospital Rastatt to the Medical Bureau XIV. Army Corps, 16.08.1916. The physician's demand was completely in line with the decree from July 1916 about the accelerated discharge of patients from hospitals, see BA-MA PHD 6/ 197: Gesichtspunkte zur Frage der beschleunigten Herausziehung militärisch oder in der Kriegswirtschaft verwendbarer Mannschaften aus den Lazaretten und Truppenteilen, 28.07.1916, 9.

25 See for example BA-MA RM 120/129: secret enactment of the Prussian War Ministry, Medical Department [*Medizinal-Abteilung*], 11.01.1916.

26 Wilmanns, *Die badischen Lazarette während des Krieges*, 5–8; Schultzen, "Organisation der zahnärztlichen Hilfe im Kriege", 126.

27 See the pre-war enactment Kriegs-Sanitäts-Ordnung (K.S.O.), 1907, § 459; this is further explained in BayKrA Stellvertretendes Generalkommando I. Armeekorps Sanitätsamt/176: Bavarian War Ministry, Medical Department, to all mobile and immobile units, 25.11.1914.

28 See, for example, BayKrA Stellvertretendes Generalkommando I. Armeekorps Sanitätsamt/176: decree by the Prussian War Ministry, Medical Department, 08.03.1915.

29 Perry, *Reycling the Disabled*, 177–179.

30 BayKrA Stellvertretendes Generalkommando I. Armeekorps Sanitätsamt/176: Bavarian War Ministry, Medical Department, to the Medical Bureaus I., II. and III. Army Corps, 20.02.1917.

31 For a discussion of the contemporary discourse on pension psychosis [*Rentensucht*] see, for example, Eghigian, "The German Welfare State as a Discourse of Trauma", 103–108; Michl, *Im Dienste des "Volkskörpers"*, 202–213.

32 Frickhinger, "Lazarettunterricht", 967.

33 Hellpach, "Lazarettdisziplin als Heilfaktor", 1209.

34 Körte, *Ueber die Versorgung unserer Verwundeten im Felde, Vortrag, gehalten am 11*, 28–32; Kolmsee, *Unter dem Zeichen des Äskulap*, 187.

35 For more on the patients' perspective and on the hospitals' fight against idleness, see Enzensberger, "Nicht nur Menschen"; concerning the hospitalisation experiences of German soldiers with facial injuries see Ruff, *Gesichter des Ersten Weltkrieges.*

36 BayKrA Stellvertretendes Generalkommando I. Armeekorps Sanitätsamt/176: Medical Bureau I. Army Corps to all reserve hospitals et al., 20.03.1917. For more on Germany's struggle over men and manpower see Feldman, *Army Industry and Labor in Germany*, 64–96.

37 BA-MA PHD 6/197: Gesichtspunkte zur Frage der beschleunigten Herausziehung militärisch oder in der Kriegswirtschaft verwendbarer Mannschaften aus den Lazaretten und Truppenteilen, 28.07.1916.

38 BA-MA PHD 6/197: Gesichtspunkte zur Frage der beschleunigten Herausziehung militärisch oder in der Kriegswirtschaft verwendbarer Mannschaften aus den Lazaretten und Truppenteilen, 28.07.1916, 16.

39 BA-MA PHD 6/197: Gesichtspunkte zur Frage der beschleunigten Herausziehung militärisch oder in der Kriegswirtschaft verwendbarer Mannschaften aus den Lazaretten und Truppenteilen, 28.07.1916, 16–17.

40 BA-MA PHD 6/197: Gesichtspunkte zur Frage der beschleunigten Herausziehung militärisch oder in der Kriegswirtschaft verwendbarer Mannschaften aus den Lazaretten und Truppenteilen, 28.07.1916, 10.

41 BA-MA PHD 6/197: Gesichtspunkte zur Frage der beschleunigten Herausziehung militärisch oder in der Kriegswirtschaft verwendbarer Mannschaften aus den Lazaretten und Truppenteilen, 28.07.1916.

42 There still exists neither a comprehensive study on nursing nor on the Red Cross in First World War Germany. For a general historic overview see Riesenberger, *Das Deutsche Rote Kreuz*; with a focus on nursing in base hospitals see Stölzle, *Kriegskrankenpflege im Ersten Weltkrieg*; for a local study on Baden Lutzer, *Der Badische Frauenverein.* 2002.

43 See for example BayKrA Stellvertretendes Generalkommando II. Armeekorps Sanitätsamt/39: Deputy General Command II. Army Corps to all reserve hospitals et al., 21.10.1914.

44 BA-MA PHD 6/197: Gesichtspunkte zur Frage der beschleunigten Herausziehung militärisch oder in der Kriegswirtschaft verwendbarer Mannschaften aus den Lazaretten und Truppenteilen, 28.07.1916, 12.

45 Frickhinger, "Lazarettunterricht," 967–968; Tabora, "Aus den Seuchenlazaretten der Festung Strassburg," 609–610.

46 See for example Hauptstaatsarchiv Stuttgart (HStAS) M 77/2 Nr. 41: report by the war medical inspector [*Kriegssanitätsinspekteur*] Lasser, 19.11.17, fol. 218, and Württembergian War Ministry, Medical Department, 20.12.15, fol. 95.

47 Eckart, *Medizin und Krieg*, 67–73.

48 Joppich, *Otto von Schjerning*, 110–112; Kolmsee, *Unter dem Zeichen des Äskulap*, 207–208.

49 Eckart and Gradmann, "Medizin," 210–219.

50 Schlich, "Welche Macht über Tod und Leben."

51 Whalen, *Bitter Wounds*, 50–53.

52 See also Perry, *Recycling the Disabled*, 93–98; Kienitz, *Beschädigte Helden*, 183–195.

53 BA-MA PHD 6/197: Gesichtspunkte zur Frage der beschleunigten Herausziehung militärisch oder in der Kriegswirtschaft verwendbarer Mannschaften aus den Lazaretten und Truppenteilen, 28.07.1916, 10.

54 See for example the documents in BayKrA Stellvertretendes Generalkommando I. Armeekorps Sanitätsamt/176, 1916–1918.

55 BA-MA PHD 6/206: Anweisung zur beschleunigten Verfügbarmachung militärisch oder in der Kriegswirtschaft verwendbarer Wehrpflichtiger und Heeresangehöriger

durch die Bezirkskommandos, Truppenteile und Lazarette (Entlassungs-Beschleunigungs-Anweisung: Eba.), 12.04.1917.

56 BA-MA PHD 6/206: Anweisung zur beschleunigten Verfügbarmachung militärisch oder in der Kriegswirtschaft verwendbarer Wehrpflichtiger und Heeresangehöriger durch die Bezirkskommandos, Truppenteile und Lazarette (Entlassungs-Beschleunigungs-Anweisung: Eba.), 12.04.1917, 5.

57 For the case of Württemberg see HStAS M 77/2 vol. 41: supplementary notes to the Eba. by the Deputy General Command XIII. Army Corps, 31.05.17, fol. 213.

58 BayKrA Stellvertretendes Generalkommando I. Armeekorps/739: Deputy General Command I. Bavarian Army Corps to all subordinated Medical Bureaus et al., 22.06.1917.

59 BayKrA Stellvertretendes Generalkommando I. Armeekorps/739: Deputy General Command I. Bavarian Army Corps to all subordinated Medical Bureaus et al., 22.06.1917.

60 For more on the "performative element" of wartime medical care see Winter, "Hospitals," 355–356.

61 See for example BayKrA Stellvertretendes Generalkommando I. Armeekorps Sanitätsamt/176: Bavarian War Ministry to the Medical Bureaus I., II., and III. Army Corps, 15.05.1917; Bavarian War Ministry to the Medical Bureaus I., II., and III. Army Corps, 13.05.1918.

62 See for example the reports in: BayKrA Stellvertretendes Generalkommando I. Armeekorps Sanitätsamt/432 and BayKrA Stellvertretendes Generalkommando I. Armeekorps/739.

63 Regarding the discussion on the totality of the First World War see the good overview by Segesser, "Controversy: Total War."

64 See HStAS M 77/2 Nr. 48: Prussian War Ministry, 04.09.15.

65 See for example HStAS M 77/2 Nr. 4: Deputy General Command XIII. Army Corps, "Denkschrift Ersatzwesen", 13–14; HStAS M 77/2 Nr. 48: Deputy General Command XIII. Army Corps, 23.01.16, fol. 16; Schultzen, "Kriegsärztliches aus Feld und Heimat", 128.

66 Rosa, *Beschleunigung. Die Veränderung der Zeitstrukturen in der Moderne*, 266.

Bibliography

Archival materials

Bayerisches Haupstaatsarchiv – Abteilung IV. Kriegsarchiv (BayKrA)

BayKrA MKr 18389.
BayKrA Stellvertretendes Generalkommando I. Armeekorps/739.
BayKrA Stellvertretendes Generalkommando I. Armeekorps Sanitätsamt/176.
BayKrA Stellvertretendes Generalkommando I. Armeekorps Sanitätsamt/432.
BayKrA Stellvertretendes Generalkommando II. Armeekorps Sanitätsamt/39.

Bundesarchiv – Militärarchiv (BA-MA)

BA-MA PHD 6/197.
BA-MA PHD 6/206.
BA-MA RM 120/129.

Hauptstaatsarchiv Stuttgart (HStAS)

HStAS M 77/2 Nr. 4.
HStAS M 77/2 Nr. 41.
HStAS M 77/2 Nr. 48.

Generallandesarchiv Karlsruhe 456 F 113 Nr. 82.
Universitätsarchiv Heidelberg H-III-600/1.

Published materials

Afflerbach, Holger. *Falkenhayn: Politisches Denken und Handeln im Kaiserreich*. München: Oldenbourg, 1994.

Bergen, Leo van. “For Soldier and State: Dual Loyalty and World War One.” *Medicine, Conflict and Survival* 28, no. 4 (2012): 291–308.

Cohen, Deborah. *The War Come Home: Disabled Veterans in Britain and Germany, 1914–1939*. Berkeley: University of California Press, 2001.

Eckart, Wolfgang and Christoph Gradmann. “Medizin.” In *Enzyklopädie Erster Weltkrieg*, edited by Gerhard Hirschfeld, Gerd Krumeich, and Irina Renz, 3rd edn, 210–219. Paderborn: Schöningh, 2009.

Eckart, Wolfgang. *Medizin und Krieg: Deutschland 1914–1924*. Paderborn: Schöningh, 2014.

Eghigian, Greg A. “The German Welfare State as a Discourse of Trauma.” In *Traumatic Pasts. History, Psychiatry, and Trauma in the Modern Age, 1870–1930*, edited by Mark S. Micale and Paul Lerner, 92–111. Cambridge: Cambridge University Press, 2010.

Ehlert, Hans, Michael Epkenhans, and Gerhard P. Groß (eds). *Der Schlieffenplan: Analysen und Dokumente*. Paderborn: Schöningh, 2006.

Enzensburger, Alina. “‘Nicht nur Menschen, sondern auch viel Zeit muß gegenwärtig totgeschlagen werden.’ Müßiggang und Mußemomente im deutschen Lazarett des Ersten Weltkriegs, 1914–1918.” *Muße: Ein Magazin*, no. 1 (2015). http://mussemagazin.de/?p=910.

Enzensberger, Alina. *Deutsche Lazarette an der Heimatfront: Militärische Planung, soldatische Kriegserfahrung und gesellschaftliche Vorstellungen eines Übergangsraums im Ersten Weltkrieg, 1914–1918 (working title)*. PhD Diss., Humboldt University of Berlin (in preparation).

Epkenhans, Michael. *Der Erste Weltkrieg*. Paderborn: Schöningh, 2015.

Feldman, Gerald Donald. *Army Industry and Labor in Germany, 1914–1918*. Princeton: Princeton University Press, 1966.

Foley, Robert T. “What’s in a Name? The Development of Strategies of Attrition on the Western Front, 1914–1918. ”*The Historian*, no. 68 (2006): 722–746.

Franklin, Benjamin. “Advice to a Young Tradesman, Written by an Old One: To My Friend A.B., 21 July 1748.” In *The Papers of Benjamin Franklin, vol. 3, 1 January 1745, through 30 June 1750*, edited by Leonard W. Labaree, 304–308. New Haven: Yale University Press, 1961.

Frickhinger, Karl. “Lazarettunterricht.” *Münchener Medizinische Wochenschrift*, no. 62 (1915): 967–968.

Groß, Gerhard P. *Mythos und Wirklichkeit: Geschichte des operativen Denkens im deutschen Heer von Moltke d.Ä. bis Heusinger*. Paderborn: Schöningh, 2012.

Harrison, Mark. "The Medicalization of War – the Militarization of Medicine." *Social Historyof Medicine*, no. 9 (1996): 267–276.

Heeres-Sanitätsinspektion des Reichskriegsministeriums. *Sanitätsbericht über das Deutsche Heer (Deutsches Feld- und Besatzungsheer) im Weltkriege 1914–1918 (Deutscher Kriegssanitätsbericht 1914–1918), vol. 3: Die Krankenbewegung bei dem Deutschen Feld- und Besatzungsheer im Weltkriege 1914/1918*. Berlin: Mittler, 1934.

Heeres-Sanitätsinspektion des Reichskriegsministeriums. *Sanitätsbericht über das Deutsche Heer (Deutsches Feld- und Besatzungsheer) im Weltkriege 1914–1918 (Deutscher Kriegssanitätsbericht 1914–1918), vol. 1: Gliederung des Heeressanitätswesens im Weltkriege 1914/1918*. Berlin: Mittler, 1935.

Hellpach, Willy. "Lazarettdisziplin als Heilfaktor." *Medizinische Klinik* 11 (1915): 1208.

Herwig, Holger H. *The Marne 1914: The Opening of World War I and the Battle that Changed the World*. New York: Random House, 2009.

Hofer, Hans-Georg and Cay-Rüdiger Prüll (eds). *War, Trauma and Medicine in Germany and Central Europe (1914–1939)*. Freiburg: Centaurus, 2011.

Joppich, Robin. *Otto von Schjerning: Wissenschaftler, Generalstabsarzt der preußischen Armee und Chef des deutschen Feldsanitätswesens im Ersten Weltkrieg*. PhD Diss., Heidelberg University, 1997.

Kienitz, Sabine. *Beschädigte Helden: Kriegsinvalidität und Körperbilder, 1914–1923*. Paderborn: Schöningh, 2008.

Kolmsee, Peter. *Unter dem Zeichen des Äskulap: Eine Einführung in die Geschichte des Militärsanitätswesens von den frühesten Anfängen bis zum Ende des Ersten Weltkrieges*. Bonn: Beta-Verlag, 1997.

Körte, Werner. *Ueber die Versorgung unserer Verwundeten im Felde, Vortrag, gehalten am 11 April 1915 im Saale des Abgeordnetenhauses*. Berlin: Hirschwald, 1915.

Kramer, Alan. "Blockade and Economic Warfare." In *Cambridge History of the First World War. Vol. 2*, edited by Jay Winter, 460–489. Cambridge: Cambridge University Press, 2014.

Kriegs-Sanitäts-Ordnung (K.S.O.) vom 27.01.1907. Berlin: Mittler, 1907.

Leonhard, Jörn. *Die Büchse der Pandora: Geschichte des Ersten Weltkrieges*. München: Beck, 2014.

Lerner, Paul. *Hysterical Men: War, Psychiatry, and the Politics of Trauma in Germany, 1890–1930*. Ithaca: Cornell University Press, 2003.

Lutzer, Kerstin. *Der Badische Frauenverein 1859–1918: Rotes Kreuz, Fürsorge und Frauenfrage*. Stuttgart: Kohlhammer, 2002.

Martineck, Otto. "Die Begriffe Dienstbrauchbarkeit (Kriegsbrauchbarkeit), Dienstbeschädigung, Erwerbsunfähigkeit, Verstümmelung im Rahmen der militärärztlichen Gutachtertätigkeit." In *Die militärärztliche Sachverständigentätigkeit auf dem Gebiete des Ersatzwesens und der militärischen Versorgung. Erster Teil*, edited by Curt Adam and Zentralkomitee für das ärztliche Fortbildungswesen in Preußen, 8–63. Jena: Gustav Fischer, 1917.

Michl, Susanne. *Im Dienste des "Volkskörpers": Deutsche und französische Ärzte im Ersten Weltkrieg*. Göttingen: Vandenhoeck & Ruprecht, 2007.

Mombauer, Annika. "German War Plans." In *War Planning 1914*, edited by Richard F. Hamilton and Holger H. Herwig, 48–79. Cambridge: Cambridge University Press, 2010.

Münkler, Herfried. *Der Große Krieg: Die Welt 1914 bis 1918*. 4th edn. Berlin: Rowohlt, 2014.

Overmans, Rüdiger. "Kriegsverluste." In *Enzyklopädie Erster Weltkrieg*, edited by Gerhard Hirschfeld, Gerd Krumeich, Irina Renz, and Markus Pöhlmann, 3rd edn. Paderborn: Schöningh, 2014.

Perry, Heather R. *Recycling the Disabled: Army, Medicine and Modernity in WWI Germany.* New York: Manchester University Press, 2014.

Prüll, Livia and Philipp Rauh. "Krieg und medikale Kultur in Deutschland, 1914–1945: Eine Einleitung." In *Krieg und medikale Kultur: Patientenschicksale und ärztliches Handeln in der Zeit der Weltkriege 1914–1945*, edited by Livia Prüll and Philipp Rauh, 7–29. Göttingen: Wallstein, 2014.

Riesenberger, Dieter. *Das Deutsche Rote Kreuz: Eine Geschichte 1864–1990*. Paderborn: Schöningh, 2001.

Rojek, Sebastian. *Versunkene Hoffnungen: Die Deutsche Marine im Umgang mit Erwartungen und Enttäuschungen, 1871–1930*. Berlin: De Gruyter Oldenbourg, 2017.

Rosa, Hartmut. *Beschleunigung: Die Veränderung der Zeitstrukturen in der Moderne.* Frankfurt am Main: Suhrkamp, 2005.

Rosa, Hartmut. *Beschleunigung und Entfremdung: Entwurf einer kritischen Theorie spätmoderner Zeitlichkeit*. Berlin: Suhrkamp, 2013.

Ruff, Melanie. *Gesichter des Ersten Weltkrieges: Alltag, Biografien und Selbstdarstellungen von gesichtsverletzten Soldaten*. Stuttgart: Steiner, 2015.

Schlich, Thomas. "'Welche Macht über Tod und Leben!' Die Etablierung der Bluttransfusion im Ersten Weltkrieg." In *Die Medizin und der Erste Weltkrieg*, edited by Wolfgang Eckart and Christoph Gradmann, 109–130. Pfaffenweiler: Centaurus, 1996.

Schmidt, Anne. *Belehrung, Propaganda, Vertrauensarbeit: Zum Wandel amtlicher Kommunikationspolitik in Deutschland 1914–1918*. Essen: Klartext, 2006.

Schultzen, Wilhelm. "Kriegsärztliches aus Feld und Heimat." *Zeitschrift für ärztliche Fortbildung*, no. 13 (1916): 126–131.

Schultzen, Wilhelm. "Organisation der zahnärztlichen Hilfe im Kriege." *Zeitschrift fürärztliche Fortbildung*, no. 13 (1916): 245–250.

Segesser, Daniel Marc. "Controversy: Total War." *1914–1918—online. International Encyclopedia of the First World War*, edited by Ute Daniel, Peter Gatrell, Oliver Janz, Heather Jones, Jennifer Keene, Alan Kramer, and Bill Nasson. Berlin: Freie Universität Berlin, 2014. http://dx.doi.org/10.15463/ie1418.10315.

Stevenson, David. *1914–1918: The History of the First World War*. London: Penguin Books, 2012.

Stölzle, Astrid. *Kriegskrankenpflege im Ersten Weltkrieg: Das Pflegepersonal der freiwilligen Krankenpflege in den Etappen des Deutschen Kaiserreichs.* Stuttgart: Steiner, 2013.

Tabora, Demeter von. "Aus den Seuchenlazaretten der Festung Strassburg." *MünchenerMedizinische Wochenschrift* no. 62 (1915): 609–610.

Ullrich, Bernd. "Krieg der Nerven, Krieg des Willens." In *Erster Weltkrieg: Kulturwissenschaftliches Handbuch*, edited by Stefan Kaufmann, Lars Koch, and Niels Werber, 232–258. Stuttgart: Metzler, 2014.

Whalen, Robert Weldon. *Bitter Wounds: German Victims of the Great War, 1914–1939*. Ithaca: Cornell University Press, 1984.

Wilmanns, Karl. *Die badischen Lazarette während des Krieges.* Berlin: Offene Worte, 1932.

Winter, Jay. "Hospitals." In *Capital Cities at War: Paris, London, Berlin, 1914–1919*, vol. 2, edited by Jay Winter and Jean-Louis Robert, 354–382. Cambridge: Cambridge University Press, 2007.

Ziemann, Benjamin. *Gewalt im Ersten Weltkrieg: Töten – Überleben – Verweigern*. Essen: Klartext, 2013.

2 Fast therapy and fast recovery

The role of time for the Italian neuropsychiatric service in the war zones

Anna Grillini

The organisation of an efficient military health service was one of the major challenges that Italian military healthcare had to face during the First World War. From the first months of combat, it was clear that the psychiatric sector had to be involved in the massive increase in health services by creating a special section: the neuropsychiatric service (*servizio neuropsichiatrico*).

Time played a key role in this service. The continuing need for men able to fight influenced, for example, the time devoted to the treatment and evacuation of the sick and wounded. Doctors had to cope with the constant pressure to diagnose and treat their patients quickly.

The psychiatrists welcomed the challenges of mobilisation with enthusiasm, and the most important members of the discipline actively contributed to the organisation of the military neuropsychiatric service. Like their colleagues, the psychiatrists also had to live with the constant pressure to arrive at quick diagnoses and rapid recoveries. Those who paid the price of this haste, on the part of the neuropsychiatric service, were the soldiers, victims of invasive therapies which aimed at sending them back to the front as soon as possible rather than making their health the top priority. The doctors themselves, who had welcomed the war as an unprecedented opportunity to observe and study new varieties of symptoms, had to deal with a military reality that undermined the value of the medical profession, while promoting discipline and dedication to the army.[1] In this conflict between professions, loyalty to the cause prevailed over the medical vocation, and the psychiatrists gradually abdicated from the role of health guardians in favour of being primarily army officers. Thus, the soldiers were perceived first of all as weapons with which to achieve victory, and were considered useful as long as they could fight or work in strategic industrial sectors.

The need to keep hospital stays as short as possible, and all the difficulties associated with the sense of urgency that characterised the psychiatric service during wartime, were more pronounced in the war-zone health structures. They were subjected to more pressure than the institutions far from the front. In particular, these issues were even more evident in the organisation of two neuropsychiatric villages, situated in Carnia and Cadore, which functioned as miniature asylums. In these places the instrumental use of time for military purposes was accentuated by spatial isolation. These two aspects – the control of time and space – combined to create a sort of limbo in which the soldiers were waiting to know their fate. So how was the military neuropsychiatric service of the Italian Army organised? In what way did the issue of time influence this particular part of the military health service? What were neuropsychiatric villages, how did they work, and what was their role in the health service? These are the main issues of this essay. The first part is dedicated to the organisation of the service, while the second analyses the psychiatric work in the villages.

The information available on the functioning and organisation of these two psychiatric units is mostly limited to what is contained in two articles, published in 1915 and 1918, and written by two of the most important Italian psychiatrists of the time, Arturo Morselli and Placido Consiglio. In addition to the information contained in the two articles, further details and guidelines on the organisation of the two neuropsychiatric villages and on the military health service are available in various reports compiled by the psychiatrist responsible for each Army. These reports were to be sent to the Italian Army High Command every three months, but most of these documents have been lost, and so a detailed reconstruction of the evolution of the psychiatric service during the conflict is difficult. However, these reports are an interesting, and hitherto neglected, testimony.

The importance of time is also evident in the published sources. The articles are from two very different periods. The first article was by Arturo Morselli, the psychiatric consultant of the First Army and a well-known physician, and published in 1915, at the very beginning of the Italian war experience. This article dealt with a new and promising medical structure that had already proved to be useful, but which would be expanded, improved, and significantly challenged with the growing number of patients over the following years. The second essay was written by one of the most active scholars of military psychiatry. Placido Consiglio was a military doctor who had already served in the Libyan War. His medical activities during the conflict were mainly focused on the problem of malingering and soldiers' mental endurance in the war.[2] Moreover, Consiglio was the director of one of the two small neuropsychiatric villages. His article from 1918 about the village which he controlled for years was published at the end of the war.

The small amount of information transmitted in these papers describes a timeless reality, a limbo where the traumatised soldiers had only one certainty, namely the inability to escape from the war. The image of a man wounded by the ugliness of the conflict was counterposed by that of the physician. For the doctors, the amount of time spent making the soldiers ready to return to active duty was directly proportional to the quality of their service to the nation, in a difficult clash between their medical duties and those as army officers. Time became synonymous with the quality of service rendered to the nation.

The organisation of the military psychiatric service and the control of time and space

At the time of Italy's entry into the war, Italian psychiatry was going through a particularly significant phase: the system of mental hospitals had spread to such an extent that 47 of the 69 provinces had an asylum, and psychiatry was finally accorded the status of an independent discipline, separating it permanently from general medicine.[3] Despite the undoubted successes of the previous decades, psychiatry was confined within the walls of asylums, and historical, social, and philanthropic ambitions were mostly scaled down to a rigidly defined professional routine within mental hospitals.[4] At the same time, the main pre-war activities of military psychiatrists (the identification and elimination of the 'different people', in other words all the young men who didn't fit into the psycho-physical standard of the army) were considered of secondary importance because of the pressing need for increasing numbers of men. They may not have embodied physical and moral perfection, but could however carry a weapon. From the outbreak of the war, quality was secondary to quantity because total war needed all the available men.[5]

In the summer of 1915, immediately after the Kingdom of Italy's entry into the war, the army laid the foundations for the organisation of the wartime neuropsychiatric service. It was organised into four different zones for the evacuation of the wounded: the first two zones were located in the areas of military operations, the first one in the immediate proximity of the front line and the second in the zone behind it. The other two areas were located farther from the front, respectively in easily reachable locations and in the rest of the country.

The Italian military health organisation drew inspiration from Russia's of 1905. The Russian organisation was conceived to foster a rapid removal of the wounded from the front line to the military hospitals in the rear. The fundamental feature of these logistics was that it was envisaged for a variable and above all boundless territory. Hence this organisation could be adapted to the varied Italian geography. The challenges that the health service faced were mainly those that today are recognised as the main characteristics of the First World War: weariness and immobility, poor hygiene, and carriers of illness. The great mobilisation of men and means involved the organisation of a health system unprecedented in Italian military history. The First World War was the first conflict in which wounds were a far greater cause of death than illness.[6]

Entry into the war saw Italian psychiatrists unanimously and unreservedly in favour of interventionism. The adherence of the medical profession was mainly realised through active participation in the organisation of the military health service. For psychiatrists, the outbreak of the First World War was mainly an opportunity to study a multitude of new symptoms and diseases observed in contexts other than traditional ones, but to do that they had to open their minds to new theories and possibilities:

> It's time that the alienists in lunatic asylums stopped believing that psychiatry is only what happens in their establishments … the illnesses of inpatients give no sense of the vastness of the field of action open to psychiatry; this is well recognised only by those who see many disturbed persons at liberty.[7]

Initially, before the baptism of fire, some doctors simply applied the behaviours and diagnostic categories used in asylums in peacetime, but soon these methods proved to be totally ineffective:

> Those who, after months of service at the front, were assigned to the observation of those who were ill in the psychic sphere, and were still unfamiliar with the psychiatric literature of war, felt a sense of discomfort and disorientation in their new daily practice: accustomed to the contact with the alienated in mental hospitals and remembering the most morbid cases more or less related to the seriousness and to the known psychopathological classification syndromes, they came up against the difficulty of classifying the mentally ill according to their acquired habit of mind and the traditional psychiatric concepts; and this difficulty left them surprised and perplexed for some time.[8]

The watershed represented by this conflict was clearly felt by Italian psychiatrists who understood the importance of an efficient system of neuropsychiatric assistance and worked with the army authorities for its realisation, managing to organise a system of transportation, field service and care to meet the challenges of modern warfare. This was one of the main merits of the medical community and its realisation was possible also thanks to the influence exerted by psychiatrists who managed to gain full recognition for the importance of their work.

Each army was provided with a psychiatric consultant.[9] These consultants had the task of visiting the shelters of their army where cases of mental or nervous illness had been reported, in order to make a diagnosis and decide on the subsequent transfer of patients. In addition, they had to organise and direct special departments ready to deal with unclear cases that required a period of observation before deciding on a possible transfer, for malingering and suspicious cases and for acute forms that required a short stay in hospital.[10]

Time had a key role in the service. These departments covered the triple function of

> ridding the front line troops of the encumbrance of those who were mentally ill and neuropathic, recognising and eliminating from among the latter the malingerers, curing the light forms of mental illness on the spot in order to send these soldiers back to their Corps, evacuating the most serious cases to the inner areas.[11]

The aims of the service included the rapid discharge of the less serious patients, the organisation of specialised wards closer to the front, and, above all, the constant observation of the most serious and persistent cases in order to discourage malingering and exaggerated symptoms.[12] The traumatised soldiers had to be removed as soon as possible from the corps. This was necessary to allow the doctors to examine the patients, formulate a diagnosis, and develop an effective therapy. The soldiers needed to be admitted to specialised departments in order to recover, at least in part, their emotional and mental stability. Moreover, the rapid removal of the traumatised soldiers from the corps was considered a priority to maintain the good health of the rest of the troops. Nervous breakdowns, shell shock, hysteria, depression, manic depressive psychosis, and, in general, all the forms of mental decline were considered dangerous for the army's stamina and were treated as epidemics to be isolated and overcome. Fear was contagious, and men had to be protected, as far as possible, from the negative influence of those soldiers who seemed to display cowardice or little perseverance in the face of warfare.[13]

The small neuropsychiatric villages

The creation of the neuropsychiatric service was the first step towards the total mobilisation of the Italian psychiatric profession. This section of the army included all psychiatric specialists and directed their efforts towards common goals, trying to reconcile the needs of medicine and those of military discipline. The work of the psychiatrists consisted of providing assistance both in combat areas – in field hospitals and specialised departments – and in military hospitals and asylums around the country. In this context, psychiatric power was exercised at its maximum through therapeutic action, which resulted in the total control of the patients.[14] The symbolic locations of this monitoring exercise were the '*small neuropsychiatric villages*', in effect miniature real asylums with almost all the characteristics of the traditional mental hospitals.

Those two 'miniature asylums' were situated in the areas immediately behind the front lines in Carnia and Cadore.[15] The perimeter of the structures was surrounded by three rows of barbed wire and monitored by armed guards. The small villages were essentially field structures, composed of tents and wooden buildings and initially organised to accommodate approximately 30 soldiers. The small village in Carnia, directed by Placido Consiglio, comprised patients' hutments, staff accommodation, a library, and three pavilions: one for meetings and lessons, another for medications and electric therapies, and the last was used as a chapel.[16] During the conflict the accommodation capacity of these structures reached between 100 and 150 units.[17]

Both the articles and the reports described these small villages as almost idyllic places, designed to allow soldiers to rest their minds. Consiglio wrote of tree-lined driveways and garden furniture and, according to Pighini's report, "these simple, rustic but elegant buildings were separated by grassy lawns and flower beds, brightened by small fountains and a drinking water fountain, by a bowling green and by rustic seats in the shade of the trees".[18] In these descriptions, only the chirping of the birds seemed to be missing, but in addition to the detailed list of the comforts provided for the sick there were also armed guards, bars on the windows, beds with restraint belts, and straitjackets.[19]

Everyday life was rigidly organised, but in these places the medical and military control was not limited to setting a routine and the containment of the movements of the hospitalised people. Here time and medical therapy were used primarily as weapons in order to unmask the malingerers and send as many men as possible back to the frontline – malingerers or not – as soon as possible. Psychiatrists welcomed new patients with caution because the suspicion that these were malingerers was always present. Consequently the days and hours were no longer used to find and apply appropriate therapies, but rather to increase the soldiers' emotional tension, leading the malingerers to reveal the truth about themselves. The tools used to achieve this were observation and therapy, which implies that the means which had always belonged to the everyday practice of medicine – to the therapeutic time – increasingly took the form of pressure to accelerate recovery – even only partial – or the admission of malingering.

Immediately after hospitalisation, the soldiers were carefully searched, deprived of their own clothes, washed, given other clothes, and put to bed. All the men followed the same procedures regardless of the specific affliction, because it enabled the agitated patients to be calmed, to rest the physically injured ones, and at the same time it guaranteed the healthcare staff greater control over and surveillance of the patients. In the first days of hospitalisation the patients were kept in strict isolation, suspended in limbo, not knowing when it would end and uncertain of the future. During the isolation period and then during the therapies the patients were carefully monitored. This allowed the healthcare staff to note any change in the clinical situation, but above all any ambiguous or suspicious attitude that could unmask the malingerers.[20]

Observation, isolation, and therapy were integral parts of the daily life of every psychiatric hospital and specialised military ward, so what were the peculiarities of the small neuropsychiatric villages? Why did the Italian military health authorities create these miniature asylums? The many issues related to these questions were all linked to the need for soldiers to return to active service and the time needed to make the traumatised soldiers able to fight again.

Among the various military and civilian health facilities where the traumatised soldiers were admitted, the small villages were characterised by a defined location and the total military context. The stay in these villages differed from the hospitalisation in the traditional military psychiatric wards because of the total isolation in which the soldiers were confined. Inside the village the only patients were those with mental problems; in the military health facilities there were different specialised wards, but here all the spaces were for the recovery of the mentally ill. Also, there were no civilian staff, and so all the social interactions were with other patients or with the military health staff. It differed particularly from asylums and psychiatric wards because the patients remained in a military environment. They could find some respite from the fatigue of trench warfare, but there was no relief from the iron military discipline. The doctors worked to maintain the soldiers' combat spirit at a high level, to encourage love for the Fatherland and enthusiasm for army life.[21] Furthermore, they worked to keep the war in the patients' daily lives, to make them understand that a transfer out of the combat zone was not an option and that the only option was to return to the corps.[22]

The creation of the psychiatric facilities so close to the front line was considered very useful to save precious time for the healthcare. On the one hand, the shocked or mentally ill soldiers needed to be helped as soon as possible. On the other hand, transporting patients to far-away health facilities, out of the war zone, took time: the time (and money) to organise the trains, the actual time to complete the trip from the front, and then the time to return from the hospital to the corps. For the army the costs, both in terms of time and money, of transporting thousands of men were not insubstantial. The reluctance to invest days, personnel, and material in such enterprises was even more pronounced when it came to the evacuation of the mentally ill. As mentioned earlier, these soldiers were first suspected of malingering or cowardice, so the military authorities' first thought was that the disease was just an excuse to leave the front. For these men, travel days could be an opportunity to improve their ability to simulate the disease or to develop better stories to justify their condition. Moreover, the doctors and the military authorities were convinced that a prolonged stay away from the front, whether on leave or for a rest, would promote discontent and the determination to simulate or exaggerate symptoms.[23] Essentially, the evacuation of the wounded was an absolutely primary aspect of the organisation of the war health service, which had to be controlled and implemented with caution. It should not be forgotten that the time invested in soldiers' transfer and hospitalisation was time taken away from the national cause, and therefore could not be wasted or used lightly. The priority was to keep the men within the war zone, where they could be controlled more easily and sent to the corps in a few hours. There they were always soldiers first, and men or patients second.

The small neuropsychiatric villages were conceived as places of first care and observation. The role of psychiatrists operating in these structures was to cure the milder cases quickly, report malingerers, and, above all, to ensure that only the most serious cases were sent behind the lines. The work in these departments was therefore dynamic and fast, characterised by a considerable turnover of patients, relatively short stays, and invasive therapies to achieve results in a short time. It could be defined as the front line of psychiatric care and, as at every first aid point, time was the key to giving better assistance. But, more than anything else, time was at the service of the army. Transport, recovery, medical observation, therapies, everything had to be fast – and not necessarily accurate – because in the end what was important was to ensure that as many soldiers as possible could hold a rifle and shoot.

The time for therapy

During the First World War, the time for therapy also assumed new and varied meanings. For the soldiers it could be a longed-for rest from the war and its violence, but for doctors it represented above all the time to make the combatants useful once again to the army, a time during which they exercised a total control over the sufferers' lives. The therapy time also represented the biggest opportunity to observe the patients' reactions to different stimuli or new kinds of therapy; it was the time when the doctors could really exercise their care and medical skills.

The control of time and space was a well-known way for psychiatrists to exercise power: it was used for the patients' recovery and to protect both society and those who were ill from reciprocal negative influences. Inspired by Philippe Pinel's theories about the 'moral treatment', the control of space and time during the therapy and recovery period became a priority. The ideal locations for care were described as isolated, clean, well ordered, and excluding the outside world, because only with these characteristics would it be possible to control the course of the disease and also every moment of the patient's life, which was regulated by a strict routine based on manual work.[24] The severe control of time was regarded as an essential element for scientific observation: the maintenance of a repetitive and precise routine was functional both for keeping an eye on a patient's reactions or symptoms and medical monitoring. This activity was to be carried out with dedication and precision because it was this work that could lead to the identification of the best treatment and then the patient's recovery. Moreover, it was a cornerstone of scientific research.[25]

In the psychiatric science of the early twentieth century, controlling the space and time of people with mental problems were fundamental elements of therapy. However, in the Italian military psychiatry of the pre-war years those two fundamental aspects were much less important. Military psychiatry was a discipline based on the elimination of 'not standard people', and it was not expected that the army physician would do more than attest a mental defect. Despite the rigid selective criteria imposed on young men, often tried by hunger and harsh living conditions, and despite the growing number of discharges, military psychiatrists periodically demanded more selective and strict rules.[26] People judged mentally unfit for military life were the responsibility of the civilian doctors and of mental hospitals. It was in these locations that effective therapies were developed and the foundations for a return to society were laid. The problems of healing and reintegration into society as productive members were issues of primary importance for the exercise of the medical profession in the civilian society, but were non-existent in the military environment. Basically, the main aspect for the practice of military psychiatry was diagnostics, endlessly exercised rather than concerning themselves with finding a cure.

The obsession with identifying individuals unsuitable for the army and the terror of a 'moral contagion' among its ranks made the military authorities myopic over the disastrous lack of preparation of military psychiatrists regarding daily psychiatric practice. Reforming and unloading unwanted people on to the shoulders of the civilian health authorities entailed an extreme specialisation of doctors in psychiatric diagnosis, but especially a widespread and deeply ingrained inability in therapy.[27]

Unfortunately, the war put both the diagnostic and therapeutic capabilities of civilian and military psychiatrists to the test, as they had to cope with the hospitalisation of more and more soldiers:

> An average of four to six soldiers were admitted daily to the camp hospitals – which normally had 20–25 beds (up to 50 or 100 in 1916) – where they remained for no more than two weeks. A survey by the Military Health Inspectorate revealed that in only three months, from May to July 1916, more than 20,000 soldiers were hospitalised for nervous and mental health disorders [...]. Due to the overwhelming number of soldiers seen, diagnoses were frequently superficial and treatment inappropriate. Often the doctors in the camp hospitals simply listed the patient's symptoms in their files, leaving the diagnosis to the next hospital.[28]

Psychiatrists working in hospitals in the war zone – as in the case of the small neuropsychiatric villages – had to decide in a short time whether soldiers needed long-term care, were mild cases, or were malingerers. In any case, all the patients had to be quickly sent to other destinations. All this made it very difficult to formulate appropriate diagnoses and treatment. Doctors were already under pressure because they faced new symptoms and were overloaded by the number of patients, but in addition to that they had to diagnose and treat the patients quickly. For traumatised soldiers, the few weeks of hospitalisation in front-line hospitals made the difference between life and death, health and permanent damage, referral to martial law or freedom, transfer to another hospital or a return to the front. In these facilities there were deadlines for transferring the sick, providing treatment, and dealing with the arrival of new patients. This last point was emphasised by the assaults. Every action in the trenches led to the hospitalisation of dozens of traumatised soldiers. Just as the soldiers at the front were getting ready to fight, the psychiatrists knew that the interval between a battle and the next one had to be optimised in order to be ready to receive a new wave of patients. Because of the lack of time, the enthusiasm that had animated the psychiatrists at the beginning of the war soon diminished. In the face of the reality of war, the hopes of physicians to make a difference and study new phenomena outside the hospital corridors were destined to be dashed: "the work was carried out at an extremely rapid pace, which clashed with their aspirations and abilities, without being able to examine the substance more thoroughly".[29]

In the small neuropsychiatric villages patients were kept under constant observation by the nursing and medical staff. During the observation period the patient was left to express his madness, with no reassurances or information about his condition. Observation was considered both a prerogative and an integral part of the therapy, which could enable the doctors to make a diagnosis. The emotional pressures and the fear of what was to follow were, in some cases, sufficient to regain at least some semblance of sanity, just enough to be sent back to the front and thus avoid painful treatments and possible allegations of malingering.

The need to achieve results in the short term and the suspicion of malingering that weighed on each mentally ill soldier resulted in the use of very invasive therapies. Among the most commonly used therapies were faradisation, hypnosis, and various techniques of suggestion. The reports drawn up by the psychiatric army consultants described how electric therapy – faradisation – was successful in 90 per cent of cases of shock mutism. The physicians applied sudden and gradual faradic currents to the sides of the patient's neck, and the discharge increased in intensity if the patient did not resume talking or stop to tremble.[30] The results achieved with electrical therapy or hypnosis were highlighted in many of the reports on the psychiatric activity of the armed forces. These techniques, and faradisation in particular, were also used on the patients considered malingerers, who therefore were theoretically perfectly healthy and not in need of treatment.

According to a study published by the medical director Gaetano Martini at the end of the conflict, the malingerers were generally young, because the elderly were less mentally agile, more family oriented, and more fearful of being reported to the military court.[31] Often these men came from the cities. They had jobs such as fishermen, street vendors, and porters. Martini also offered a general physical description of these soldiers:

> The mouth is half open, the nasolabial folds barely evident, the gaze is sometimes fixed, at other times wandering aimlessly. Sometimes the eyes are exaggeratedly open, and the forehead creased to mimic the state of fright. It is not difficult to surprise the malingerer looking furtively and duly directing his gaze especially at the examiner, trying to know the impression produced, the final judgment. The eyes maintain their normal luster. Sometimes, spontaneously or with some harsh reproach, especially in novices, redness appears on the face, a sign of shame and embarrassment.[32]

Throughout the duration of the conflict, finding malingerers was an obsession of military doctors, a war within the war.[33] The role of doctors in treating the malingerers was the ultimate realisation of the double goal of the medical-military profession, including therapy and discipline: "Doctors, in fact, not only had to discover whether a soldier was a malingerer: they had a nobler task, to return a soldier to the Army, a citizen to the Fatherland."[34] The psychiatrists of the villages had to identify the malingerers and had to do it in the short time available to avoid the evacuation from the front of even one of these fake mentally ill soldiers. After the observation period, the doctors tried to persuade the suspected malingerers to admit their guilt and return to do their duty as combatants but, if the persuasions did not work, they started with more coercive methods such as intimidations and threats. Domenico Isola, a doctor, described the rapidity and effectiveness of the electrical therapy. According to Isola, the patient who was truly ill cooperated right away, while the malingerer found every excuse not to engage in therapy and did not make any progress. In these cases, the electric therapy enabled the doctors to identify the malingerer: by increasing the current to intolerable levels, the impostor soon confessed in order to stop the pain.[35] The time required to obtain results with electrical therapy varied from patient to patient, but each session could last more than an hour and therapy could be performed several times per week. In the report on psychiatric assistance in the 3rd Army, it was reported that, as a result of careful observation and therapies such as faradisation, the time within which to decide a soldier's fate decreased from 15–30 days to 8–15. With a 50 per cent saving on the time normally expected, doctors tried to reduce overcrowding and send soldiers back to the front.[36]

The hospitalisation in the psychiatric villages was based on the fundamental assumption that the return to the front was not a question of 'if' but 'when'. The awareness of the inevitability of this event had to be instilled in the consciousness of every soldier.[37] For this reason, as well as for logistical reasons, the facilities were located near the front, positioned so that the muffled noise of the fighting reached the hospitalised soldiers as a reminder of the inevitability of their return to war.[38] The time spent in the villages was therefore characterised by the suspicion of being a malingerer with which each soldier was welcomed, by several therapies – more functional for the army than for the patient – and by the constant contact with the conflict:

> The therapeutic process lost its positive, enlightening approach, which had always been the basis of its strategies, and it became yet another weapon in the hands of a military health service less interested in relieving the suffering of the sick than in discouraging the malingerers who, when declaring their madness, were believed to be wanting to escape from the dangers of the battlefield. Here the condemnation of the lunatic in the forced hospitalisation immediately behind the front lines was regarded as therapy.[39]

The violent sensorial stimulation together with the proximity to the fighting, the emotional pressure due to the doctors' suspicions, and the uncertainty that accompanied the hospitalisation combined to create a sort of limbo from which the only way out seemed to be to return to the front or to fall into an ever-deeper psychosis. These pressures were considered integral parts of the therapeutic process that thus absorbed every minute of the day without being limited to very specific times. The patient was completely immersed in an environment, temporal as well as spatial, designed to have the maximum therapeutic impact in the shortest possible time. Every minute and every space was dedicated to a therapeutic function. Even leisure became essential to maintain morale and, above all, the men's patriotism.[40]

The application of those therapies that we might call 'punitive' and deeply invasive, in order to recover the combatants rather than the men, could be considered the result of the difficulty of reconciling the medical and military vocations. In this conflict of professionalism the loyalty to the army and to the national cause prevailed over the duty to care for the sick, thus reducing the soldiers to mere tools for achieving victory.

At the end of the period of hospitalisation in the small villages the soldiers could return to the front or continue their healthcare, obtaining the status of being sick and removing themselves, at least in part, from the suspicion of being malingerers. If the combatants were in need of prolonged therapy they were transferred to psychiatric facilities behind the lines. The observation and care did not last for more than 90 days, after which either they were reintegrated into the military corps or, less desirably for the military authorities, it might be possible to begin the procedure for discharge or a period of further treatment.[41]

With the outbreak of the war, the increasing demands for men to send to the front led to a loss of any positive value of therapy and of the control with which it was applied, in a context of total mobilisation where the search for mental health was secondary to using therapy to send the soldiers back to the trenches.[42] With these premises, the importance of time meant reducing the number of days the soldiers remained far from the front, or pressuring a malingerer to commit a faux pas. Even the time needed to transfer the soldiers to hospitals behind the lines was a matter of concern. Time, or rather the lack of it, was a reality with which the military doctors had to deal: it affected the daily therapy and the doctor–patient relationship.

Notes

1 Bianchi, *La follia e la fuga*, 58–84; Crouthamel and Leese, *Psychological Trauma and the Legacies of the First World War.*
2 Scartabellati, "Un Wanderer dell'anormalità", 89–112.
3 Bianchi, *La follia e la fuga*, 65–66; Giovannini, "La psichiatria italiana e la Grande Guerra", 111–152.
4 Stock, *La formazione della psichiatria*, 142–143.
5 Consiglio, "Studi di psichiatria militare", 370.
6 Biagini and Gionfrida, "L'organizzazione della sanità militare italiana al fronte nella prima guerra mondiale", 214–223; Cadeddu, 2009, 77–88; De Napoli, *La sanità militare in Italia durante la I guerra mondiale*, 20–24; Vagnini, "Il Servizio neuro-psichiatrico nella sanità militare italiana nella Grande Guerra", 225–236.
7 Morselli, "Review of F. Boucherat, Des maladies mentales dans l'armée en temps de guerre", 236.

8 De Lisi and Foscarini, “Psiconevrosi di guerra e piccole cause emotive”, 14–15.
9 For the 1st Army it was Arturo Morselli, for the 2nd Army Vincenzo Bianchi, for the 3rd Army Angelo Alberti, and Giacomo Pighini for the 4th Army.
10 Tamburini, “L’organizzazione del servizio neuro”, 178–179; Tamburini and Antonini, *L’assistenza degli alienati in Italia e nelle varie nazioni*, 670–675.
11 Tamburini, “L’organizzazione del servizio neuro”, 178–179.
12 The importance of the objectives being pursued and the growing number of soldiers hospitalised for nervous and mental health disorders showed the need for more military psychiatrists. For these reasons, a psychiatric training unit was established at the camp university at San Giorgio di Nogaro (Udine) in 1916. The number and use of military psychiatrists were at the centre of quite heated discussions between the president of the Associazione dei Medici dei Manicomi Italiani (Association of Italian Asylum Physicians), Raffaele Brugia, and the Army General Consultant for Neuropsychiatric Issues, Augusto Tamburini. See: Brugia, “Lettera del direttore”; Tamburini, “Sul servizio psichiatrico di Guerra”, 509.
13 Antonini, “La questione della epurazione dall’esercito”, 17–25; Funaioli, “Organizzazione del servizio medico-psichiatrico di guerra”, 338; Pighini, “Per la eliminazione di degenerati psichici dall’esercito combattente”, 978–996.
14 Foucault, *Le pouvoir psychiatrique*, 15–21.
15 Tamburini and Antonini, *L’assistenza degli alienati in Italia e nelle varie nazioni*, 661.
16 Consiglio, “Un villaggio neuro-psichiatrico in Zona di Guerra”, 175.
17 Tamburini, “L’organizzazione del servizio neuro psichiatrico di guerra nel nostro esercito”, 179–180.
18 Archive of the Army Staff Historical Office, henceforth AASHO, E7, b. 31, 1915.
19 Tamburini, “L’organizzazione del servizio neuro psichiatrico di guerra nel nostro esercito”, 174; Morselli, “Il reparto neuro-psichiatrico dell’ospedale da campo di 100 letti 032 (III Armata)”, 390.
20 Manente and Scartabellati, “Gli psichiatri alla guerra. Organizzazione militare e servizio bellico, 1911–1919”, 108–109.
21 Consiglio, “Un villaggio neuro-psichiatrico in Zona di Guerra”, 12.
22 AASHO, E7, b. 31, *Relazione trimestrale sul Servizio Psichiatrico*, secondo trimestre 1917, 1.
23 AASHO, E7, b. 31, 1918.
24 Foucault, *Le pouvoir psychiatrique*, 14–16.
25 Pinel, *Traité médico-philosophiques sur l’aliénation mentale ou la manie*, 95–96.
26 Bucciante, “Profilassi della delinquenza nell’esercito”, 248–259.
27 Scartabellati, “Un Wanderer dell’anormalità”, 92–93.
28 Bianchi, “Psychiatrists, Soldiers, and Officers in Italy During the Great War”, 227–228.
29 Alberti, “Le psicosi di Guerra”, 269.
30 AASHO, E7, b. 31, 1917.
31 Martini, La simulazione della pazzia nei militari, 232.
32 Martini, La simulazione della pazzia nei militari, 232.
33 Gibelli, *L’officina della guerra*, 154.
34 Gibelli, *L’officina della guerra*, 155.
35 Gibelli, *L’officina della guerra*, 155.
36 AASHO, E7, b. 31, 1917.
37 Manente and Scartabellati, “Gli psichiatri alla Guerra: Organizzazione militare e servizio bellico, 1911–1919”, 104–107.
38 Consiglio, “Un villaggio neuro-psichiatrico in Zona di Guerra”, 173–174.
39 Manente and Scartabellati, “Gli psichiatri alla Guerra: Organizzazione militare e servizio bellico, 1911–1919”, 108–109.
40 Consiglio, “Un villaggio neuro-psichiatrico in Zona di Guerra”, 174.
41 AASHO, E7, b. 31, 1917.

42 Manente and Scartabellati, "Gli psichiatri alla Guerra: Organizzazione militare e servizio bellico, 1911–1919", 108–109; Scartabellati, "Il dovere dei medici nell'ora presente", 65–94.

Bibliography

Unpublished materials

Archive of the Army Staff Historical Office (AASHO)

AASHO, E7, b. 31, *Relazione sul servizio neuropsichiatrico della IV armata dal 1 gennaio al 31 giugno*, 1915.

AASHO, E7, b. 31, *Relazione sul funzionamento del servizio psichiatrico della 3 Armata*, secondo semestre 1917.

AASHO, E7, b. 31, *Relazione trimestrale sul Servizio Psichiatrico*, secondo trimestre 1917.

AASHO, E7, b. 31, *Relazione sul funzionamento della consulenza delle malattie nervosa durante il trimestre luglio – settembre 1918 (3 Armata).*

Published materials

Alberti, A. "Le psicosi di guerra." In *Atti del I Convegno Nazionale per l'assistenza agli invalidi di Guerra*. Milan: Koschitz, 1919.

Antonini, Giuseppe. "La questione della epurazione dall'esercito." *Archivio di AntropologiaCriminale, Psichiatria e Medicina Legale*, no. 38(1917): 17–40.

Biagini, Antonello and Gionfrida, Alessandro. "L'organizzazione della sanità militare italiana al fronte nella prima guerra mondiale." In *In bona salute de anima e de corpo: Malati, medici e guaritori nel divenire della storia*, edited by Giuseppe Motta, 214–223. Milan: Ed. Franco Angeli, 2007.

Bianchi, Bianca. *La follia e la fuga: Nevrosi di guerra, diserzione e disobbedienza nell'esercito italiano 1915–1918*. Rome: Bulzoni, 2001.

Bianchi, Bianca. "Psychiatrists, Soldiers, and Officers in Italy During the Great War." In *Traumatic Pasts: History, Psychiatry and Trauma in the Modern Age, 1870–1930*, edited by Mark S. Micale and Peter Lerner, 223–252. Cambridge: Cambridge University Press, 2001.

Brugia, Raffaele. "Lettera del direttore." *Bollettino dell'Associazione tra i Medici dei-Manicomi pubblici italiani*, no. 5(1915): 10.

Bucciante, Angelo. "Profilassi della delinquenza nell'esercito." *Rivista Sperimentale diFreniatria* 36, no. 3(1912): 248–259.

Cadeddu, Lorenzo. "La sanità militare nella Grande Guerra." In *Malattie e medicina durante la Grande Guerra, 1915–1919*, edited by Nicola Bettiol, Ernesto Brunetta, Daniele Ceschin, De Bertolis, Lucio Fabi, Ambrogio Fassina, and Daniele Toffolon, 77–88. Udine: Gaspari editore, 2009.

Consiglio, Placido. "Studi di psichiatria militare." *Rivista Sperimentale di Freniatria* 36, no. 3(1912): 370–412.

Consiglio, Placido. "Un villaggio neuro-psichiatrico in Zona di Guerra." *Rivista Sperimentaledi Freniatria* 42, no. 2(1918): 175–182.

Crouthamel, Jason and Peter Leese. *Psychological Trauma and the Legacies of the First World War*. Basingstoke: Palgrave Macmillan, 2017.

De Lisi, Lionello and Ezio Foscarini. "Psiconevrosi di guerra e piccole cause emotive." *Note eriviste di psichiatria* 1 (1920): 10–58.

De Napoli, Domenico. *La sanità militare in Italia durante la I guerra mondiale.* Rome: Ed. Apes, 1989.

Foucault, Michel. *Le pouvoir psychiatrique: Cours au Collège de France 1973–1974.* Seuil: Gallimard, new edn2004.

Funaioli, Gaetano. "Organizzazione del servizio medico-psichiatrico di guerra." *RivistaSperimentale di Freniatria* 29, no. 4(1911): 337–368.

Gibelli, Antonio. *L'officina della Guerra: La Grande Guerra e le trasformazioni del mondo mentale.* Turin: Universale Bollati Boringhieri, 2007.

Giovannini, Paolo. "La psichiatria italiana e la Grande Guerra: Ideologia e terapia psichiatrica alle prese con la nuova realtà bellica." *Sanità, scienza e storia*, no. 1 (1978): 111–152.

Manente, Silvia and Andrea Scartabellati. "Gli psichiatri alla guerra: Organizzazione militare e servizio bellico, 1911–1919." In *Dalle trincee al manicomio. Esperienza bellica e destino di matti e psichiatri nella Grande guerra*, edited by Andrea Scartabellati, 91–119. Turin: Marco Valerio, 2008.

Martini, Giuseppe. "La simulazione della pazzia nei militari." *Rivista Sperimentale diFreniatria* 44, no. 2(1919): 228–250.

Morselli, Arturo. "Il reparto neuro-psichiatrico dell'ospedale da campo di 100 letti 032 (III Armata)." *Quaderni di Psichiatria*, no. 2(1915): 390–410.

Morselli, Arturo. "Review of F. Boucherat, Des maladies mentales dans l'armée en temps de guerre." *Quaderni di psichiatria*, no. 3(1916): 236.

Pighini, Giacomo. "Per la eliminazione di degenerati psichici dall'esercito combattente." *Giornale di medicina militare*, no. 2(1918): 978–996.

Pinel, Philippe. *Traité médico-philosophiques sur l'aliénation mentale ou la manie.* Paris: Richard, Caille et Ravier, 1800.

Scartabellati, Andrea. "Un Wanderer dell'anormalità? Un invito allo studio di Placido Consiglio (1877–1959)." *Rivista Sperimentale di Freniatria* 3(2010): 89–112.

Stock, Fabio. *La formazione della psichiatria.* Rome: Il pensiero scientifico, 1981.

Tamburini, Augusto. "L'organizzazione del servizio neuro psichiatrico di guerra nel nostro esercito." *Rivista Sperimentale di Freniatria*, 1(1916): 165–182.

Tamburini, Augusto. "Sul servizio psichiatrico di Guerra." *Rivista Sperimentale di Freniatria*, 1(1916): 509.

Tamburini, Ferrari and Giuseppe Antonini. *L'assistenza degli alienati in Italia e nelle varie nazioni.* Turin: Utet, 1918.

Vagnini, Alessandro. "Il Servizio neuro-psichiatrico nella sanità militare italiana nella Grande Guerra." In *In bona salute de anima e de corpo: Malati, medici e guaritori nel divenire della storia*, edited by Giuseppe Motta, 225–236. Milan: Ed. Franco Angeli, 2007.

3 A stitch in time

Inefficiency and the appeal of patriotic work in Australia and Canada

Steve Marti

The First World War marked the first substantial military mobilisation for the Dominion of Canada and the Commonwealth of Australia, the largest and most populous of Britain's settler colonies. The outbreak of war in 1914 sparked an enthusiastic response from these two self-governing colonies, henceforth referred to as dominions, which both produced a considerable outpouring of voluntary contributions to the imperial war effort. Nearly 13 per cent of military-aged males in Australia and Canada enlisted, and their casualty rates were staggering. The Canadian Expeditionary Force (CEF) suffered a 50 per cent casualty rate, while two-thirds of Australian soldiers were killed or wounded over the course of the war.[1] Most notably, these contingents were sustained largely through voluntary enlistment – exclusively so in the case of the Australian Imperial Force (AIF).

The dominions, still fledgling states in 1914, relied heavily on voluntary contributions to transform their skeletal peacetime armies into such formidable expeditionary forces. Alongside calls for enlistments, the voluntary mobilisation of the dominion home fronts produced substantial material contributions to the national and imperial war effort. All 23 aeroplanes of the 1st Squadron, Australian Flying Corps, were purchased by popular subscription campaigns.[2] In Canada, the Imperial Order Daughters of the Empire (IODE) – Canada's largest women's voluntary society – collected funds to equip 36 hospital wards and provided 19 motor ambulances, 22 sterilising units, and 942 hospital cots over the course of the war.[3] The willing contribution of these funds, not to mention thousands of human-hours of voluntary labour, are suggestive of the broader 'totalizing logic' of First World War mobilisation. As John Horne and other historians point out, however, tensions soon developed between spontaneous grassroots efforts that joined in the work of national mobilisation and the state's effort to coordinate these disparate initiatives into a more efficient war effort.[4]

State and society maintained relatively cordial relations in Australia and Canada, largely because state authorities passed few measures to regulate voluntary mobilisation. Wartime measures policed civilian morale through censorship and guarded against sedition with enhanced powers of arrest and detention, but the dominion governments did not pass measures that would enable the state to take a stronger role in coordinating voluntary patriotic work, such as the enactment of the Director General of Voluntary Organisation and the National Council of Social Services to oversee and coordinate patriotic work in Britain.[5] Canada's Wartime Charities Act only required charities to register with state authorities and submit to financial oversight. Australian states established State War Councils, but these lacked the authority to compel disparate groups of volunteers to band together into a coordinated effort. In the absence of stronger legislation, the coordination of voluntary action in the dominions nevertheless raised tensions between the competing interests of local, regional, national, and imperial war efforts, and revealed how geographic boundaries divided the process of imperial mobilisation into smaller efforts.[6] The constraints of time proved just as important as space in shaping the negotiations between state and society over the coordination of the voluntary war effort.

Space presented a formidable obstacle in Australia and Canada, which were separated by oceans from both the imperial metropole as well as the theatres of war. The voluntary war effort, however, could alter perceptions of space. Robert Rutherdale and Ian Miller examine patriotic festivals on the Canadian home front to demonstrate how such performances brought the war home for those who stayed behind while friends and family departed for service overseas.[7] Joining a knitting circle or a sewing group turned patriotic work into an 'emotional labour' that provided volunteers with an opportunity to fulfil gender or class roles, to socialise, or to cope with the separation or death of loved ones serving overseas.[8] Patriotic donations could also project peripheral communities to the front lines. Donors who purchased machine guns to be donated to the CEF engraved these gifts with inscriptions such as "Abbotsford District" or "Gun to be returned to Sandon if in existence at termination of war".[9] The Springsure Red Cross purchased a motor ambulance and requested that the ambulance be labelled "Springsure Queensland", and asked that "any part of it that is left" at the end of the war be returned to the town so it could be put on display outside the shire hall.[10] Through these requests, local communal efforts projected contributions from the furthest corners of the Empire and, in some cases, attempted to bring a part of the war back to their hometown. Such contributions helped bridge the gaps between local mobilisation schemes and the war overseas. Voluntary mobilisation also made tangible the imagined connections that drew the disparate corners of the Empire together. As one Montreal member of the Imperial Order Daughters of the Empire observed of her organisation's wartime work, "in knitting a Balaclava cap we were knitting the Bond of Empire".[11] Participating in the war effort could bridge the gap between the metropole and the periphery. In his comparative study of mobilisation in Northampton and Béziers, Pierre Purseigle highlights the confluence of local motivations with larger ideas of nationalism and patriotism.[12] While voluntary mobilisation could shrink the imagined space between the home front and the front lines, the time taken to ship donated goods overseas provided a cogent reminder that geographical distances remained constant.

As patriotic work brought the war home, civilians on the dominion home fronts became increasingly familiar with the conditions of industrialised warfare. Descriptions of life in the trenches and reports of the lethality of machine guns, artillery, aerial combat, tanks, and poison gas reached home in newspapers and soldiers' letters. The voluntary mobilisation of the dominions aimed to provide their soldiers with the materials to overcome these hardships. Volunteers knitted socks to help stave off trench foot, sewed bandages to mend shrapnel and gunshot wounds, assembled respirators to provide protection from poison gas, and raised money to purchase a machine gun or aeroplane. Producing sewed or knitted items by hand, in a social setting such as a knitting circle, may have been emotionally rewarding for volunteers, but this method of providing comforts to soldiers overseas proved time-consuming and extended delays along the lengthy supply lines between the home front and the front lines. The moral and material benefits of voluntary mobilisation pulled in opposite directions.

This chapter examines the correspondence generated by the voluntary and patriotic societies in Australia and Canada to reveal how time factored into the effort to mobilise human and material resources across space. Voluntary mobilisation might close the imaginary spaces between the home front and the front line, as well as between the imperial metropole and disparate peripheral communities of the dominions. The distance between Australia or Canada and the theatres of war presented a challenge for dominion defence authorities, coordinating the transportation of human and material resources across the world. The geographical distances remained constant, but a cohesive war effort could make better use of shipping space and reduce the time needed to send aid from the home front to the front lines. The dominion governments' reluctance to take a stronger role in coordinating the voluntary war effort revealed how the emotional labour of voluntary patriotic work weighed against the material needs of the war effort.

The outbreak of war and the imminent departure of dominion expeditionary forces created a sudden urgency to contribute to the war effort. The threat of German raiders at large in the South Pacific and the Atlantic required dominion contingents to sail in convoy. Once the first contingent departed, it could be months before another sailed. Donors raced to provide some comfort or sustenance to soldiers before the first contingents sailed, and contributions of food, clothing, and equipment came in as quickly as volunteers rushed to enlist in August 1914. Canadian farmers offered part of their crops to help feed the soldiers of the First Contingent mustering at Valcartier. Victor Sinclair, a barrister from Tilsonburg, Ontario wrote to Canada's Minister of Militia, Sir Sam Hughes, on behalf of local farmers who offered their harvest of winter apples to the expeditionary force, if the Department of Militia could wait three weeks while the rest of the crop was prepared for storage. The contingent, however, was scheduled to depart before the donation could reach the soldiers.[13] Elliot G. Stevenson wrote on behalf of the Independent Order of Foresters of Toronto to donate apples from the orchard maintained by their Orphans' Home in Oakville, but only if the Department of Militia could "see to the freight or expenses".[14] When it was discovered that Camp Valcartier was closed, Stevenson informed the Director of Supplies and Transport, Colonel J. Lyon Biggar, that the Independent Order of Foresters had no means of storing the apples for a subsequent contingent, and would sell the apples at market value if the Department of Militia did not pay the freight to bring them to where they were needed. The Department of Militia agreed to ship the apples, but due to an error they were delivered to Captain Wilson of Toronto's Salvation Army and distributed among the urban poor, rather than to soldiers of the CEF.[15]

Australians likewise responded to the outbreak of war with hurried offers of support, eager for Australian soldiers to know their generosity before they embarked. A register kept by the Australian Department of Defence lists almost two hundred separate donations of cash and kind made in the opening months of the war. The items varied from a barrister's offer to draft soldiers' wills free of charge, to "a quantity of papaw ointment".[16] Donations of foodstuffs, particularly those with a longer shelf life, were readily received while some gifts, such as the use of 500 camels and their drivers from Abdul Wade of Wangamana, New South Wales, were dismissed as "not required at present".[17] Other donations seemed useful but were difficult to accept. Gifts of livestock presented a particular challenge for defence authorities, which did not possess the facilities or manpower to tend these animals or to slaughter and process the carcasses into rations. Melbourne barrister E.J. Cordner brokered a donation of 100 sheep on behalf of his clients, but the Department of Defence simply sold these and added £118 to its coffers.[18] The Premier of South Australia presented the Defence Department with a gift of 500 sheep. South Australian drovers donated these sheep intending to feed volunteers from their own state, but the gift was declined unless the sheep were slaughtered and shipped overseas at the donors' expense.[19] Defence authorities also received offers to provide soldiers with necessary equipment. A.E. Gadd of Nagambie, Victoria offered a pair of binoculars, on the condition they were returned at the cessation of hostilities, while Miss A.M.A. Gibson donated her binoculars but was asked to have them repaired before they could be accepted.[20]

These hurried donations were offered in the spirit of patriotism, but this haste left gaps between what donors could provide and what military authorities required. Donors offered surplus foodstuffs, equipment, or services, but these gifts were sometimes offered with the expectation that the state would help cover expenses or at least allow the donor to recoup some lost profit. The reliance on the state to provide the necessary funds to turn a patriotic gesture into a viable contribution to the war effort revealed part of the inefficiency of voluntary contributions. As dominion expeditionary forces made urgent preparations during the Boreal autumn of 1914, civilians likewise rushed to the colours and offered whatever contributions they could in order to take their part in this hurried mobilisation. The scramble of 1914 settled into an uneasy routine, as patriotic societies formed communal hubs to orchestrate the voluntary mobilisation of the dominions. Gathering around the nuclei of existing voluntary societies, women joined local knitting circles and men organised collections to purchase military equipment. Patriotic work settled into a regular routine, but donors did their best to ensure their work kept pace with changing conditions at the front.

The desire to improve the condition of soldiers overseas motivated volunteers on the home fronts to participate in patriotic work and contribute to the overall war effort. Stories from the front lines filtered back home with details of the horrid condition soldiers faced overseas. In reading these accounts, volunteers sprang into action to correct whatever material deficiencies exacerbated the soldiers' hardship. The value of these contributions depended on whether they could reach soldiers in time. As Canada's First Contingent faced cold, wet conditions encamped on the Salisbury Plain over the fall and winter of 1914, Elizabeth Evans wrote on behalf of the Home Workers of Quebec City offering to supply wood-burning stoves for the soldiers. Eugène Fiset, Deputy Minister of Defence, did his best to explain that the proposed donation was not practical because the Canadian Contingent would soon cross the Channel to take up camp in France or Belgium. While the stoves seemed well suited to the situation at Salisbury, they might not be needed by the time they caught up with Canadian soldiers. Nor could the contingent simply store the stoves until necessary, as Fiset explained, because "the road to glory cannot be followed with much baggage".[21] The offer of stoves seemed like a perfectly rational and practical solution to address the conditions endured by Canadian soldiers training on the Salisbury Plain, but the ever-changing conditions in which Canadian soldiers would be billeted led the Department of Militia to reject the offer.

When dominion soldiers entered sustained combat operations in April 1915, descriptions of trench warfare convinced many at home of a desperate shortage of sandbags. In Sydney, the executives of the 19th Battalion Comforts Club debated whether they should devote their efforts to making sandbags, respirators, or fly-veils for Australian soldiers fighting at Gallipoli. The members decided that the shortage of sandbags presented the most pressing need for soldiers at the front.[22] In Melbourne, the Lady Mayoress's Patriotic League made a public appeal to collect empty 70-pound sugar sacs, so they could be shipped to the front and repurposed as sandbags.[23] Rather than collect sugar sacs, the 19th Battalion Comfort Club set to work sewing sandbags soon after the start of the Gallipoli Campaign, and the effort to produce sandbags spread to other parts of Australia over the next few months. An appeal from the Western Australia Sand Bag Committee prompted the citizens of Williams, Western Australia to form the Williams District War Emergency League, whose first action was to organise a dance in aid of the Western Australian Sand Bag Fund Day appeal.[24] Mrs Lloyd Williams, Honorary Secretary of the Queensland Sand Bag League, pleaded for the state premier's attendance at their next meeting, in the hopes that the residents of Brisbane could do "equally well" as the effort in Victoria and New South Wales, who were "producing bags by the hundred thousands [sic]".[25] The perceived need for sandbags at Anzac Cove prompted a surge in voluntary work dedicated to alleviating these shortages. The call for volunteers reached Williams District in mid-November, while the Queensland Sand Bag League beseeched the Premier for his support during the last days of December. Whether any sandbags from these later efforts reached Gallipoli before the last Allied troops withdrew during the first week of January 1916 is doubtful.

The initiatives to provide soldiers overseas with specific equipment such as stoves or sandbags to correct a particular shortage at the front reflected the engagement of volunteers on the home front, who sought to provide a meaningful contribution to improve the condition of soldiers overseas. Newswires delivered reports from dominion contingents overseas to the home front in a matter of days, and donors rushed to alleviate an acute shortage of equipment at the front. The desire to provide a useful contribution to the war effort motivated these generous offers, but volunteers proved slow to coordinate contributions that could keep pace with the changing conditions at the front. Just as with the rush to war in August and September 1914, the delays in producing and shipping materials to soldiers overseas could create some friction between state and society, while the inability to deliver these materials to soldiers in time resulted in wasted effort.

The distance between the home front and the front lines made it difficult for volunteers to respond effectively to changing conditions on the battlefield, but voluntary societies could easily anticipate changing seasonal conditions and major holidays. The demand for winter clothing, or the desire to offer soldiers an extra treat at Christmas, injected the voluntary war effort with renewed enthusiasm. The Australian Red Cross found themselves in an ideal position to respond to the onset of winter in 1914. The New South Wales Division proudly reported that their recent consignment of sheepskin vests, mittens, and wool blankets had been marked with labels advertising their place of manufacture: Liverpool, New South Wales. For British soldiers in the trenches, these comforts would "show what Australia can do with her wool".[26] In July 1917, Red Cross Branches in Victoria met to coordinate their efforts for the remainder of the Austral winter. The Victorian executives of the Red Cross stressed the urgency of using the next three months to produce warm woollens and flannel underwear that could be delivered to Australian soldiers in France in time for soldiers facing the onset of the Boreal winter.[27] Canadian voluntary societies likewise anticipated the need for winter clothing, but collected these in the Boreal summer months of June and July. The Local Council of Women in Halifax encouraged members to use their summer holidays preparing woollen comforts for the Red Cross, so they could reach Canadian soldiers in the trenches in time for winter.[28] The changing of seasons continued to punctuate the onslaught of volunteer work. In April 1916, the South Australian branch of the League of Loyal Women reported that the collection of blankets and rugs had ended, the demand for gloves and waistcoats had subsided, and that its members should switch to making shirts out of khaki twill.[29]

Demands for winter clothing destined for soldiers in France and Belgium coincided with the collection of Christmas parcels. The predictable timing of this annual celebration required a carefully timed appeal for donations so that Christmas parcels could be assembled, packed, and shipped to reach soldiers by 25 December. By 1916 and 1917, Canadian voluntary societies became familiar enough with shipping times to begin preparing Christmas parcels in September and October. The Milltown Soldiers' Comforts Association, on the New Brunswick border with the United States, put out their call for Christmas supplies in late September and set a deadline to ship these packages by 10 October.[30] More ambitious appeals needed additional time for preparation. The Municipal Chapter of the IODE in Ottawa set out to collect and fill 10,000 Christmas stockings from its branches in the Ottawa Valley, and began making appeals as early as August.[31] The South Australian Depot of the Australian Flying Corps Comforts Club likewise kept to a strict schedule in preparing Christmas shipments. For Christmas 1917, boxes would be packed by 29 August, checked over and ready for shipment by 5 September.[32]

Christmas parcels flooded the donation boxes of voluntary societies. The South Australian division of the Trench Comforts Fund reported that their 1917 drive for Christmas parcels accounted for over one third of the year's donations.[33] Accommodating this influx took some additional care in coordinating shipments. Requiring donors to prepare parcels of standard size made it easier to pack these into shipping crates, maximising the quantity shipped while minimising the shipping space required. Volunteers with the Australian Flying Corps Comforts Club spent a week inspecting Christmas parcels to ensure they conformed to the required dimensions.[34] The Australian Red Cross, which coordinated one of the largest efforts to ship Christmas parcels, sent out sample boxes to its sub-branches so that members knew what to pack and, just as importantly, how to pack their items into a box that would fit into a larger shipment full of Christmas parcels.[35] Voluntary societies poured a disproportionately large amount of work into the collection and preparation of parcels celebrating a single day of the year, but this annual effort added to the seasonal variation of voluntary work by creating an intense, acute appeal for patriotic contributions.

Much like those efforts to address hardships encountered by soldiers at the front, the effort to send parcels to soldiers overseas in time for Christmas caught the attention of donors and volunteers at home. In preparation for Christmas 1915, the Montreal Athletic Association raised funds to send Christmas parcels to approximately 300 members of the association serving overseas. While the members of the association had contributed money to various patriotic collections during the previous year, the ladies' committee responsible for assembling and packing the Christmas parcels found "great pleasure" in this work.[36] Despite not receiving any acknowledgement that the members of the association received any of these parcels, the women who assembled the Christmas packages found the act of preparing parcels for soldiers so rewarding that they quickly formed a new committee to produce items such as pyjamas, towels, and bandages for the Red Cross. This committee intended to send these gifts to No. 3 Canadian General Hospital, formed by the Faculty of Medicine at McGill University, but offered their contributions to whichever Canadian hospital needed these items the most.[37] While members of the Montreal Athletic Association had donated money for the war effort, the occasion of Christmas 1915 – the first Christmas that Canadian soldiers spent in the trenches – prompted members of the association to contribute more than just money. The work of preparing Christmas parcels proved so rewarding that the members decided to devote themselves to producing comforts for Canadian soldiers in hospital.

Episodes of urgency punctuated the voluntary mobilisation of the dominion home fronts, but soldiers in the trenches required a constant supply of comforts, such as socks and other necessities. Donors and volunteers rallied enthusiastically to fill pressing demands to provide supplies to dominion contingents before they sailed, to correct an acute shortage of equipment, or to send parcels to soldiers in time for Christmas, and the constraints of time could raise tensions between volunteers and state or military authorities. In the rush to deliver food and supplies to the expeditionary force in August and September 1914, donors provided apples or sheep to feed soldiers but required the defence forces to transport or slaughter these before they could be used. The effort to provide Canadian soldiers with stoves or Australian soldiers with sandbags caught the attention of volunteers at home, but lingering doubts about whether these contributions could reach dominion soldiers in time placed military authorities in the awkward position of refusing donations. These hurried donations revealed how the slow and inconsistent outputs of the voluntary war effort often placed more stress on dominion supply lines, while offering contributions of questionable military value. The tensions created by these episodes of urgency were not the result of any attempt by the state to impose order on wartime society.

The constant need to provide soldiers overseas with comforts and necessities presented volunteers and military authorities with a different set of time constraints. Care parcels could be delivered at any time, unlike a Christmas package, yet donors still grappled with the challenges of producing and packaging these parcels quickly while military authorities struggled to deliver comforts to soldiers overseas in a timely fashion. The time and cost of shipping gifts such as foodstuffs, knitted comforts, or locally purchased equipment remained a constant problem for voluntary societies and military authorities in Australia and Canada. The German campaigns of unrestricted submarine warfare in 1915 and 1917 exacerbated the shipping shortage and further raised the cost of sending goods – particularly from Australia. Despite different attempts to coax the voluntary war effort into adopting measures that would reduce shipping times and make better use of shipping space, the insistence on producing hand-made comforts for soldiers overseas reveals how the work of donors and volunteers on the home front continued with little regard for the constraints of time.

Despite placing a heavier burden on the dominions' supply lines, donation of material goods remained the preferred method of contributing to the war effort. The coordinators of voluntary societies such as the Red Cross noted the drawbacks of shipping material goods from Australia and Canada to the front. In June 1915, Canadian Red Cross workers based in London wrote to donors at home requesting donations of cash, rather than donations in kind. Colonel Charles Hodgetts, who oversaw the Canadian Red Cross's activities in London, argued that donations of money presented the most economical means of sending aid to soldiers or prisoners of war. Rather than sending knitted comforts, financial contributions also gave the Red Cross the flexibility to purchase goods in Britain in response to the changing needs at the front, or in prisoner camps, and eliminated the costs of packaging and trans-oceanic shipping.[38]

Voluntary societies continued to raise substantial sums of money over the course of the war, but volunteers found it difficult to abandon the work of knitting and sewing and devote themselves to contributions of cash. At a meeting of representatives from 64 branches of the Australian Red Cross in Victoria, delegates argued over the best means to provide comforts to Australian prisoners of war. The Commissioner of the Red Cross Depot in Melbourne, W.F. Greenwood, proposed sending a mix of knitted comforts and cash donations, arguing that knitted comforts could be stored at the depot until shipping space became available. The representative from the Geelong branch argued that Red Cross work should focus on fundraising. Besides taking up valuable shipping space, moths could destroy woollen items as they sat in a warehouse or the hold of a ship. These practical considerations could not overrule the moral incentive of making comforts by hand. By the meeting's end, no consensus could be reached, and individual branches decided for themselves whether to contribute funds or knitted items.[39]

Even the members of local branches debated between themselves whether to devote their efforts to knitting items or sending funds overseas. A member of the Dufferin Chapter of the IODE, in Vancouver, noted that wool could be acquired more cheaply in England than Canada. The chapter could purchase a greater quantity of socks in England than its members could produce by knitting socks with locally sourced wool purchased at the equivalent value. This member went on to suggest disbanding the chapter's weekly knitting circle so that members could concentrate on raising funds to purchase socks and other comforts from suppliers in Britain. While the members of the chapter resolved to send some of their available funds "to England or where most needed", they did not stop knitting. At the next month's meeting the chapter's 16 members turned in 19 pairs of socks and voted to spend $75 of the chapter's funds to purchase more wool for their knitting circle.[40]

Patriotic societies organised public solicitations to collect money for the Red Cross and other patriotic funds, but the members of these societies also retained money from these collections to subsidise their knitting circles. In October 1918, the women of the Dufferin Chapter, IODE allocated $200 of their collected funds for the Canadian War Contingent Association and the Canadian Red Cross's fund for prisoners of war. Having made these donations, the members of the Dufferin Chapter voted to spend $25 to purchase wool for further knitting projects.[41] The Milltown Soldiers' Comfort Association raised funds on behalf of the Red Cross and the St John Ambulance, but its members also wanted to knit socks for soldiers overseas. While preparing a shipment of parcels in June 1918, the members discovered that they did not have enough socks prepared to fill all their parcels. Despite having money on hand to purchase the necessary socks, the members decided to delay sending this shipment by a week so that more socks could be knitted.[42] In Australia, the Fitzroy Branch of the Red Cross responded to an urgent appeal to provide clothing and comforts for Australian soldiers and set out to raise £500 to meet these needs but, like their counterparts in Canada, some of these funds were used to purchase sewing or knitting supplies for their members to make some of the clothing themselves.[43] The personal fulfilment gained by knitting socks or sewing items of clothing for soldiers overseas provided volunteers more satisfaction than sending donations of money to fund the purchase of materials in Britain. The act of preparing a personal contribution for friends, family, or associates motivated volunteers to continue donating their time and energy to the war effort.

The tension between providing donations of cash or kind revealed a paradox in the voluntary mobilisation of the dominions. By all measures, monetary donations proved to be the most efficient means of contributing to the war, but the satisfaction of taking part in the manufacture of home-made comforts motivated volunteers and sustained their willingness to continue contributing to the work of wartime mobilisation. Not only did these gifts take up valuable shipping space, the volunteers who knitted socks or sewed comforts in their spare time could not match the speed or scale of Britain's textile industry. Some volunteers pointed out that socks and other comforts could be purchased in greater quantities and more rapidly in Britain than they could be knitted or sewn at home, but the act of producing a tangible contribution to the war effort provided a sense of emotional fulfilment to volunteers coping with the separation or loss of loved ones in wartime. The production of clothing and other hand-made items through voluntary labour on the Canadian and Australian home fronts was a slow and costly method of contributions for the war effort, yet the communal act of knitting or sewing collectively remained the preferred method for volunteers on the home front to contribute to the war.

Just as knitting together helped forge communal bonds at home, sending parcels overseas maintained bonds with friends and loved ones serving overseas. While the distance between the dominions and the theatres of operation in Europe and the Middle East theatres raised the cost of sending parcels overseas, the task of delivering parcels to specific soldiers or units only complicated the work of keeping soldiers supplied over the tangled lines of logistics that stretched from the home front to the front lines. Patriotic societies such as the Australian Comforts fund or the Canadian War Contingent Association worked with their respective departments of defence to coordinate the collective shipment of care packages. These organisations shipped parcels free of charge, but reserved the right to distribute these packages based on the needs at the front, rather than the request of the sender. The Australian Comforts fund and the Canadian War Contingent Association presented a sensible compromise between the desire to send comforts to soldiers overseas and the need to streamline dominion supply lines, yet the delivery of comforts to soldiers remained a deeply contentions issue for donors at home.

Sending a parcel through the Australian Comforts fund or the Canadian War Contingent Association was efficient but impersonal. The relatives of soldiers serving overseas preferred to form their own battalion comfort clubs to send gifts to members of a particular unit. The South Australian chapter of the League of Loyal Women began coordinating work for the state's branch of the Australian Trench Comforts Fund in late 1915. In mid-1916, the League responded to requests for separate regimental comforts clubs, and by October volunteers had organised ten regimental comforts clubs in South Australia. These clubs attracted nearly 100 additional volunteers to the work of the Trench Comforts Fund.[44] The Victoria League administered the Trench Comforts Fund in Western Australia and found that their decision to form battalion comforts clubs in November 1916 led to an eight-fold increase in donations over the following year.[45]

Forming battalion or regimental comforts clubs appealed to donors and volunteers at home, but ensuring that these parcels reached the right unit could be difficult, given the scale and complexity of the imperial supply system. Donors remained adamant that gifts reached their intended recipients. The Nova Scotia Steel Company demanded that the representative for Canada's Department of Militia in London, Colonel John Wallace Carson, confirm that the 50 pounds of tobacco donated by the company had indeed been distributed to soldiers from Nova Scotia.[46] Daniel Chrisholm, a commissioner of the City of Toronto, enquired about the distribution of 14,000 packages of cookies that the municipal government sent as a Christmas gift for soldiers from Toronto. Chrisholm's enquiry was prompted when the Mayor received a letter of thanks from a soldier in the British Army, which raised suspicion that at least some of the 14,000 Toronto soldiers had not received their packet of biscuits.[47] Crawford Vaughan, the Premier of South Australia, enquired about a donation of sheep made in December 1914, intended to benefit soldiers from South Australia. The sheep were slaughtered and shipped to England, as instructed by the Department of Defence, but could not be re-routed to follow the AIF to Egypt and were sold on delivery in England, so that the profits from the sale could purchase comforts for the benefit of South Australian soldiers. When they were unable to account for these subsequent purchases, Vaughan chastised the Defence Department and suggested it would be "advisable that the donors should be consulted as to the disposal of the amount realised".[48]

The organisation of regimental comforts clubs attracted volunteers to the League of Loyal Women's Trench Comfort Fund in South Australia, but donors also complained when parcels did not reach their intended recipients. The 18th Battery Club in Adelaide complained to the State War Council that their parcels were not reaching the men of the 18th Battery, and threatened to break from the Trench Comforts Fund. After some investigation, the State War Council determined that the Club's parcels arrived in Egypt after the 18th Battery had been transferred to the Western Front and, rather than reroute the 18th Battery's parcels to Europe, they were distributed among soldiers in the Light Horse.[49] Others found that neither the Trench Comforts Fund nor the regimental comforts clubs could meet their needs. Mr J.H. Woods complained to the State War Council that the Trench Comforts Fund had yet to send even a single cigarette to soldiers from his town of Little Swamp.[50] Miss E.R. Schramm requested permission to raise funds to send comforts to soldiers from Little Swamp, but the Council insisted Schramm work through the Trench Comforts Fund and promised to investigate the delivery of parcels to soldiers from Little Swamp.[51]

Disputes over the distribution of parcels even caused a rift between the different divisions of the Australian Red Cross. The Red Cross instructed donors to comply with standardised measurements when packing Christmas parcels, so that more parcels could fit into a single shipment without wasting space. Consolidating all Red Cross Christmas parcels in Australia into a single shipment promised to simplify the logistics of sending these gifts overseas, but this initiative required state and regional branches to send their parcels to Melbourne. Giving the Victorian Division control over the distribution of all Red Cross Christmas parcels raised concerns among some branches. In early 1917, the Queensland Division of the Red Cross decided to prepare their own shipment of Christmas parcels – independent of the Australia-wide effort coordinated in Melbourne. The Queensland Division wished to determine which hospitals would receive Christmas parcels packaged in their state, but the executives in Brisbane still requested an example of the standard Red Cross parcel to ensure that all patients received gifts of equivalent value as other Australian Red Cross parcels. While recognising the importance of uniformity, the Queensland Division nevertheless decided to go its own way to send Christmas parcels overseas.[52]

The delivery of comforts to soldiers overseas further demonstrated how the motives of donors and volunteers on the home front prolonged the delivery of supplies to dominions soldiers at the front. The dominions' supply systems worked hard to provide comforts to soldiers at the front, but complaints registered over missing deliveries of cookies, tobacco, or sheep revealed the difficulty of delivering parcels to specific men or units, which seldom remained in one place for long. Despite these difficulties, battalion or regimental comforts clubs organised on the promise that donors and volunteers could direct parcels to the members of a single unit attracted more support from donors and volunteers. These clubs contributed more funds and resources to the war effort, but the splintering of comfort funds worked against the effort to consolidate parcels and ease the burden on dominion supply lines. The scheme that proved most effective in attracting contributions of comforts and care parcels also exacerbated the delays in delivering these comforts to soldiers overseas.

To examine Australian and Canadian wartime mobilisation solely in terms of quantity suggests the dominions' state and society cooperated relatively successfully in negotiating the totalizing logic that placed an increasing amount of human and material resources in support of the imperial war effort. Traditional social structures proved their resilience, as so much of the war effort fell under the direction of voluntary societies. The state, meanwhile, passed compulsory measures such as taxation and policed wartime dissent but, compared to European belligerents, state authorities in the dominions remained remarkably reluctant to intervene upon the autonomy of voluntary initiatives. The easy relationship between state and society in the coordination of wartime mobilisation nevertheless produced prodigious donations of volunteer hours, funds, and other resources. Considering how this quantity of resources was mobilised over space and time, however, reveals a more disjointed relationship between state and society. For all the enthusiasm displayed on the dominion home fronts, knitting circles and regimental comforts clubs remained slow and inefficient methods for delivering the products of the voluntary war effort to the front lines.

The correspondence generated by the voluntary war effort reveals that the coordination of patriotic work was driven as much by the needs of soldiers overseas as it was by the motives of donors and volunteers at home. While historians have pointed out that the emotional labour of patriotic work could close the imagined distance between the home front and the front lines, the imperial periphery and the metropole, the constraints of time remained constant throughout the process of voluntary mobilisation. When time constraints pressed tighter, the relationship between volunteers and military authorities grew tense. Donors offered foodstuffs during the rush to war in 1914, but expected military authorities to coordinate its transport to training bases or embarkation points. Volunteers worked to relieve immediate needs at the front, but defence authorities had to turn away efforts that did not keep with changing conditions at the front. Even when time was short, donors and volunteers expected defence authorities to accommodate the pace of voluntary mobilisation. The attempts to streamline the delivery of parcels ran against volunteers' demands to form smaller comforts clubs, while donors protested when their parcels were diverted because of operational constraints. These tensions revealed that voluntary mobilisation remained slow and unwieldy.

While military authorities struggled to hasten voluntary efforts, the slower and more costly work of knitting, sewing, or packing personal comforts appealed to donors. Even when volunteers collected monetary donations, they preferred to spend those funds on knitting supplies to make socks themselves, rather than wire the money so that socks could be purchased overseas. Participating in the collective effort of voluntary mobilisation constituted an important social act that motivated the emotional labour of patriotic work. These self-directed efforts exacerbated delays along dominion supply lines, but ultimately appealed to volunteers and attracted more generous donations. Focusing on time, as well as space, in studying production and delivery of comforts to dominion soldiers reveals that the moral gains of the voluntary war effort counted more than material production.

While state authorities could not shorten the distance separating the dominions from the theatres of war, an efficient war effort could reduce the time necessary to produce or deliver materials from the home front to the front lines. The distance that presented such a logistical challenge for dominion military authorities, however, also protected the dominion home fronts from the military threats that loomed over European belligerents such as Britain, France, Germany, or Italy. Donors and volunteers mobilised enthusiastically to assist with the imperial war effort but, given the safety of space and the luxury of time, state authorities did not need to risk exacerbating tension with their constituents to hasten the delivery of voluntary contributions to the theatres of operation. Voluntary mobilisation moved at the pace of the home front, not the front lines. Donors and volunteers in Australia and Canada participated enthusiastically in the process of wartime mobilisation, so long as they could contribute in their own time.

Notes

1 Darwin, *The Empire Project*, 333–334.
2 List of presentation aeroplanes, 1914–1919. AWM25.
3 Pickles, *Female Imperialism and National Identity*, 21.
4 Horne, "Introduction", 3–7.
5 Grant, *Philanthropy and Voluntary Action in the First World War.*
6 Marti, "For Kin and County"; Marti, "'One big fund'".
7 Miller, *Our Glory and Our Grief*, 198–200; Rutherdale, *Hometown Horizons*, 46–88.
8 Beaumont, "Whatever Happened to Patriotic Women, 1914–1918?"; Scates, "The Unknown Sock Knitter"; Ziino, "At Home with the War".
9 "Machine Guns." LAC, RG 24.
10 Letter from Frances C. McLean to P.J. McDermott, 26 October 1915. Queensland State Archives.
11 Municipal Chapter of Montreal Minute Book, 24 September 1914. Bibliothèque et archives nationales du Québec.
12 Purseigle, "Beyond and Below the Nations".
13 Letter from Victor Sinclair to Sam Hughes, 11 September 1914. LAC, RG 24.
14 Letter from E.G. Stevenson to Sam Hughes, 6 Oct 1914. LAC, RG 24.
15 Letter from Capt. A. Bell to Col J. Lyon Biggar, 30 November 1914. LAC, RG 24.

16 Register of gifts to assist the war effort arranged by military district, Line 87. AWM 27.
17 Register of gifts to assist the war effort arranged by military district, Line 104. AWM 27.
18 Register of gifts to assist the war effort arranged by military district, Line 88. AWM 27.
19 Register of gifts to assist the war effort arranged by military district, Line 170. AWM 27.
20 Register of gifts to assist the war effort arranged by military district, Line 141. AWM 27.
21 Letter from Eugène Fiset to Elizabeth Evans, 3 November 1914. LAC, RG 24.
22 19th Battalion Comforts Fund Executive Committee Minute Book, 7 July 1915. Mitchell Library.
23 Undated notice about drive to collect sugar-bags for use as sandbags in Gallipoli. Public Records Office of Victoria.
24 Williams District War Emergency League Minute Book, 12 November 1915; 17 November 1915. J.S. Battye Library.
25 Letter from Mrs Lloyd Williams to Chief Secretary, Premier's Department, 29 December 1915. Queensland State Archives.
26 First Annual Report, 30 November 1914. N.S.W. Division, Australian Red Cross Society.
27 Victoria Division of the Red Cross Minute Book, 10 July 1917. Archives of the Red Cross of Victoria.
28 Halifax Local Council of Women Minute Book, 15 Jul 1915. Provincial Archives of Nova Scotia.
29 League of Loyal Women Minute Book A, 5 April 1916. State Library of South Australia.
30 Milltown Soldiers' Comforts Association Minute Book, 25 September 1915. Provincial Archives of New Brunswick.
31 Wales Chapter Minute Book, 13 August 1918. Archives of Ontario.
32 Australian Flying Corps Comforts Fund Minute Book, 1 August 1917. State Library of South Australia.
33 *Second Report: 1917–1918*, 6.
34 Australian Flying Corps Comforts Fund Minute Book, 1 August 1917. State Library of South Australia.
35 Executive Committee Minute Book of the Red Cross Society Queensland Division, 24 April 1917. John Oxley Library.
36 Letter from Montreal Athletic Association to John Carson, 16 November 1915. LAC RG 9-III.
37 Letter from Montreal Athletic Association to John Carson, 28 December 1915. LAC RG 9-III.
38 Letter from Mrs. Hogget to Prof. Robertson 13 June 1915, LAC MG 28-I35; Letter from Evelyn Buckeley to Mrs Herridge 15 June 1915, LAC MG 28-I35.
39 Victoria Division of the Red Cross Minute Book, 10 July 1917. Archives of the Red Cross of Victoria.
40 Dufferin Chapter Minute Book, 2 June 1915; 8 July 1915. City of Vancouver Archives.
41 Dufferin Chapter Minute Book, 8 October 1918. City of Vancouver Archives.
42 Milltown Soldiers' Comforts Association Minute Book, 4 June 1918. Provincial Archives of New Brunswick.
43 Letter from C.R.E.T. Apps to Alexander Bell, 23 February 1918. Public Records Office of Victoria.
44 League of Loyal Women Minute Book B, 16 October 1916. State Library of South Australia.
45 Report for year ending December 31, 1917, Trench Comforts Fund Victoria League. J.S. Battye Library of Western Australian History.
46 Letter from Col. J.W. Carson to LCol Charles F. Winter, 11 May 1915. LAC RG9-A-1.
47 Letter from Daniel Chrisholm to General Wood, 6 January 1916. LAC RG9-A-1.

48 Letter from Crawford Vaughn to PMO, 30 June 1915. National Archives of Australia.
49 Letter from Victor Ryan to Secretary of Central Trench Comforts Club, 4 May 1917. State Records of South Australia.
50 State War Council of South Australia Minute Book, 4 March 1918. State Records of South Australia.
51 State War Council of South Australia Minute Book, 15 January 1918. State Records of South Australia.
52 Executive Committee Minute Book of the Red Cross Society Queensland Division, 10 April 1917. John Oxley Library.

Bibliography

Archival sources

19th Battalion Comforts Fund Executive Committee Minute Book. MitchellLibrary. MLMSS 178, 19th Battalion Comforts Fund records, 1915–1919.

Australian Flying Corps Comforts Fund Minute Book. StateLibrary of South Australia. SRG 287, Australian Flying Corps Comforts Fund, South Australian Depot. Minute Book, May 1916–July 1919.

Dufferin Chapter Minute Book. City of Vancouver Archives. M255, Daughters of the Empire Fonds. AM 515-A-5, File 2 Dufferin Chapter Minute Book, June 1910–January 1916.

Executive Committee Minute Book of the Red Cross Society Queensland Division. JohnOxley Library. OM.BH Red Cross Society, Queensland Division. 1, Executive Committee Minute Books. 2, Minute Book, 1917–1919.

Financial Records Valcartier Chapter. City of Vancouver Archives. M255, Daughters of the Empire Fonds. AM 556-A-6, File 4 Valcartier Chapter Financial records, 1912–1918.

First Annual Report, 30 November 1914. Mitchell Library. Q355.85 N.S.W. Division, Australian Red Cross Society. A, Annual report.

Halifax Local Council of Women Minute Book. ProvincialArchives of Nova Scotia. MG 20, Local Council of Women. Vol 535, No. 6, Minute book, November 1913–November 1916.

League of Loyal Women Minute Book A. StateLibrary of South Australia. SRG 684, League of Loyal Women. Series 1, Vol 1, Minutes, Committee MeetingsAugust 1915–October 1916.

League of Loyal Women Minute Book B. StateLibrary of South Australia. Council Meeting, 16 October 1916. SRG 684, League of Loyal Women. Series 1, Vol 2, Minutes, Committee Meetings October 1916–October 1922.

Letter from Capt. A. Bell to Col J. Lyon Biggar, 30 November 1914. Library and Archives Canada. RG 24, Department of National Defence. HQ 54-21-33-4, Offer of apples to Overseas Contingent by Independent Order of Foresters.

Letter from Col. J.W. Carson to LCol Charles F Winter, 11 May 1915. Library and Archives Canada. RG9-A-1, Department of National Defence. Vol. 7. 4–1-3, Tobacco Sent by NS Steel Co. for 1st Division.

Letter from Crawford Vaughn to PMO, 30 June 1915. National Archives of Australia. Prime Minister's Office. A2, Correspondence files, annual single number series. 1915/3920. Meat – Gift for South Australian Troops in Egypt.

Letter from C.R.E.T. Apps to Alexander Bell, 23 February 1918. Public Records Office of Victoria. VPRS 2500/P0000/110, City of Ballarat, General Correspondence Files.

Letter from Daniel Chrisholm to General Wood, 6 January 1916. Library and Archives Canada. RG9-A-1, Department of National Defence. Vol. 7. 4–1-11. Christmas Gifts for Troops Cookies for Toronto soldiers.

Letter from E.G. Stevenson to Sam Hughes, 6 October 1914. Library and Archives Canada. RG 24, Department of National Defence. Vol 1036. H-Q 54-21-33-1, The Farmers of South Oxford Offer oats and apples to Contingent.

Letter from Eugène Fiset to Elizabeth Evans, 3 November 1914. Library and Archives Canada. RG 24, Department of National Defence. Vol 1036. HQ 54-21-33-9, Elizabeth Evans offers to supply through 'The Home Workers' Quebec Giant Heaters for use of Overseas Troops European War.

Letter from Evelyn Buckeley to Mrs Herridge 15 June 1915. Library and Archives Canada. MG 28-I35, War Work. Vol 4. File 4–15, World War I Canadian Red Cross Correspondence.

Letter from Frances C. McLean to P.J. McDermott, 26 October 1915. Queensland State Archives. Agency 112, Chief Secretary's Department. Series 5384, Batch and Miscellaneous Subject Files. Item ID861783, Batch file Correspondence and papers re Great War, 1914–1918.

Letter from G. McMilken to T.J. Ryan, 15 November 1915. Queensland State Archives. Agency 112, Chief Secretary's Department. Series 5384, Batch and Miscellaneous Subject Files. Item ID861783, Batch file Correspondence and papers re Great War, 1914–1918.

Letter from Montreal Athletic Association to John Carson, 16 November 1915. Library and Archives Canada. RG 9-III, Ministry of the Overseas Military Forces of Canada. File 4–1-11, Christmas Gifts for Troops.

Letter from Montreal Athletic Association to John Carson, 28 December 1915. Library and Archives Canada. RG 9-III, Ministry of the Overseas Military Forces of Canada. File 4–1-13, [Unnamed file].

Letter from Mrs. Hogget to Prof. Robertson 13 June 1915. Library and Archives Canada. MG 28-I35, War Work. Vol 4. File 4–15. World War I Canadian Red Cross Correspondence.

Letter from Mrs. Lloyd Williams to T.J. Ryan, 29 December 1915. Queensland State Archives. Agency 112, Chief Secretary's Department. Series 5384, Batch and Miscellaneous Subject Files. Item ID861783, Batch file Correspondence and papers re Great War, 1914–1918.

Letter from Victor Ryan to Secretary of Central Trench Comforts Club, 4 May 1917. State Records of South Australia. GRG32, State War Council of South Australia. 1/00003, Correspondence, 1917.

Letter from Victor Sinclair to Sam Hughes, 11 September 1914. Library and Archives Canada. RG 24, Department of National Defence. HQ 54-21-33-4, Offer of apples to Overseas Contingent by Independent Order of Foresters.

List of presentation aeroplanes, 1914–1919. Australian War Memorial. AWM25, Written records, 1914–1918 War. Box 375. File 3, Correspondence regarding gifts of Money; Motor Ambulances by Darling Downs District of Queensland; Lady Hamilton Recreation Hut, Tel-el-Kebir; [See also AWM25 item 375/6] Gift of Cinema Plant for each Division from Australian Comfort Funds; Gift of 12 HP Adley car; Motor cars and bicycles donated by the people of Australia.

"Machine Guns." Library and Archives Canada. RG 24, Department of National Defence. Vol. 1847. HQ 54-21-33-55, Inscriptions on Donated Military Equipment.

Malaspasia (Kitsilano) Chapter Minute Book. City of Vancouver Archives. IODE-Fonds. AM 515-A-1 file 4, Malaspasia (Kitsilano) Chapter Minute Books, May 1913–December 1917.

Milltown Soldiers' Comforts Association Minute Book. ProvincialArchives of New Brunswick. MC415, Milltown Soldiers' Comforts Association, 1917. MS1A Minute Book 1917–1940.

Municipal Chapter of Montreal Minute Book. Bibliothèque et archives nationales du Québec. P678, Fonds IODE. S3.SS1.D5, Municipal Chapter of Montreal. 2/28, Procès verbaux et rapports annuels 1910–1916.

Queensland War Council Minute Book. QueenslandState Archives. Series 16748, Queensland War Council Minute Books. Item 314694, Minute Book 2–9/10/1916–29/1/1917.

Register of gifts to assist the war effort arranged by military district. AustralianWar Memorial. AWM 27, Records arranged according to AWM Library subject classification. Box 576, File 3.

Report for year ending December 31, 1917, Trench Comforts Fund Victoria League. J. S. Battye Library of Western Australian History. PR1663, Trench Comforts Fund Victoria League.

State War Council of South Australia Minute Book. StateRecords of South Australia. GRG32 State War Council of South Australia. 3/00001, Minute books, 1916–1939.

Undated notice about drive to collect sugar-bags for use as sandbags in Gallipoli. PublicRecords Office of Victoria. VPRS 16668, Subject Correspondence Files. P0001/15, Recruiting Campaign 1916.

Victoria Division of the Red Cross Minute Book. Archives of the Red Cross of Victoria. V09, Box 1. Divisional Council Minutes. Division and Executive Committee, 1917.

Wales Chapter Minute Book. Archives of Ontario. F806, Imperial Order Daughters of the Empire (Wales Chapter) fonds. Minute Book, 1915–1924.

Williams District War Emergency League Minute Book. J.S. BattyeLibrary. MN 2417. Williams District War Emergency League. Minute book, 1915–1945.

Published primary sources

Second Report, 1917–1918: Trench Comforts Fund of the League of Loyal Women, South Australian Division, Australian Comforts Fund. Adelaide: W.K. Thomas & Co, 1918.

Published secondary sources

Beaumont, Joan. "Whatever Happened to Patriotic Women, 1914–1918?" *Australian Historical Studies* 115(2000): 284–285.

Darwin, John. *The Empire Project: The Rise and Fall of the British World-System, 1830–1970*. Cambridge: Cambridge University Press, 2009.

Grant, Peter. *Philanthropy and Voluntary Action in the First World War: Mobilizing Charity.* New York: Routledge, 2014.

Horne, John. "Introduction." In *State, Society, and Mobilization in Europe during the First World War*, edited by John Horne, 3–7. Cambridge: Cambridge University Press, 1997.

Marti, Steve. "For Kin and County: Scale, Identity, and English-Canadian Voluntary Societies, 1914–1918." *Histoire sociale/Social History* 47, no. 94(2014): 333–351.

Marti, Steve. "'One big fund': The Struggle to Centralise Australia's Voluntary War Effort, 1914–1918." *History Australia* 13, no. 3(2016): 368–381.

Miller, Ian Hugh Maclean. *Our Glory and Our Grief: Torontonians and the Great War.* Toronto: University of Toronto Press, 2002.

Pickles, Katie. *Female Imperialism and National Identity: The Imperial Order Daughters of the Empire.* Manchester: Manchester University Press, 2002.

Purseigle, Pierre. "Beyond and Below the Nations: Towards a Comparative History of Local Communities at War." In *Uncovered Fields: Perspectives in First World War Studies,* edited by Jenny Macleod and Pierre Purseigle, 95–123. Boston: Brill Academic Publisher, 2004.

Rutherdale, Robert Allen. *Hometown Horizons: Local Responses to Canada's Great War.* Vancouver: UBC Press, 2004.

Scates, Bruce. "The Unknown Sock Knitter: Voluntary Work, Emotional Labour, Bereavement and the Great War." *Labour History* 81(2001): 29–49.

Ziino, Bart. "At Home with the War: The Great War in Victorian Private Life." *Victorian Historical Journal* 86, no. 1(2015): 7–25.

4 Slow going

Wartime affect and small press modernism

Cedric Van Dijck[1]

Listing his daily activities in the trenches, Guillaume Apollinaire's short poem "4h." sees the poet get dressed in the morning only to become "an invisible being", as the colour of his French army uniform blends with the blue skies.[2] Virginia Woolf, in turn, writes of an air raid on London that woke her and forced her to evacuate downstairs to the kitchen: "one talks through the noise, rather bored by having to talk at 5 A.M. than anything else [...]. Then silence. Cocoa was brewed for us, & off we went again."[3] Both offhand remarks speak of a life in wartime conditioned by routine movements; the war is presented as slow, uneventful. In *Ugly Feelings*, Sianne Ngai shows how such low affective states, while unsuitable "for forceful or unambiguous action", nevertheless hold the power to "diagnose situations, and situations marked by blocked or thwarted action in particular".[4] Indeed, Apollinaire's feeling of being interchangeable bears an imperfect resemblance to Woolf's own irritation with restrictions placed on her movements.

In this chapter, I ask how initiatives in printing came into view in response to this curiously dual affective sensation of tedium blended with the precariousness of existence. "I don't know how much is fear, how much is boredom," Woolf wrote after another Gotha air raid on London.[5] Such a low feeling, I argue here, can entail its own kind of productivity: in her comment above she goes on to note that "[t]oday we have printed, & discussed the raid, which, according to the Star I bought was the work of 25 Gothas".[6] While the parallel drawn between her work at the press and the tense historical moment is left implicit, Woolf's words give a sense of how printing helped contemporaries recover a sense of agency. First, this chapter looks at how the notion of self became foggy and conflicted in a protracted war, as it traces the affect of boredom through the fictional archive of 1914–1918. Within this affective context, it then explores Apollinaire's involvement in the trench press in 1915, before turning to the launch of the Hogarth Press in 1917 in a final section. The chapter's emphasis on the war's 'ugly feelings' rather than its patriotic passions, such as anger and pride, necessitates, as Jeffrey Auerbach confirms, a shift away from "public records and the popular press and [towards] records intended to be private".[7] Apollinaire's "4h." was written on the back of a ripped army catalogue; Woolf's account of the air raid is an entry from her diary.

There are many ideological and aesthetic differences separating Woolf from Apollinaire. Yet this chapter sets up a dialogue between them to underline modernism's investment in print culture, and to show that that investment need not be premised on modernity's clichéd logic of speed. Modernist private presses existed in large numbers and ranged from modest operations, such as John Middleton Murry's short-lived Heron Press and Nancy Cunard's experimentation with antique printing, to large-scale enterprises, including the Yeats family's Cuala Press (1908–1946) in Dublin, Three Mountains (1923–1926) and Shakespeare and Company (1922–1929) in Paris, and the Poetry Bookshop (1913–1926) in London. Scholarship on the phenomenon, however, often fails to take the war into much consideration. In an exemplary fashion, J.H. Willis Jr. argues that the conflict "acted against presses and publishers", singling out as two important stages in the development of private presses the years between the late nineteenth century and 1914, and the 1920s.[8] Furthermore, as by-products of the technological advance of modernity, these ventures, putting out little magazines and enabling the creation of international movements and networks, figured centrally in the making and disseminating of modernism, and it is in this way that they have been studied.[9] This chapter, by contrast, highlights the much slower material practicalities of seeing work to print – typesetting, inking, binding, posting – and views this process as a tangible way of coping with some of the pressures particular to modern warfare.

Thick time

In what is now his best-known poem, the working-class poet Isaac Rosenberg describes the break of day in the trenches:

> The darkness crumbles away
> It is the same old druid Time as ever,
> Only a live thing leaps my hand,
> A queer sardonic rat,
> As I pull the parapet's poppy
> To stick behind my ear [...][10]

Consider Rosenberg's careful choice of verb with which he opens the action: "crumble away". The darkness does not "break", as its title suggests, but here fades at a more unhurried pace; dawn "begins to grow", as Wilfred Owen put it elsewhere.[11] For now, no man's land remains a "sleeping green between", not yet stirred to life by the plosive sounds ("pull the parapet's poppy") that echo gunfire to come or by the shell ("the hiss, the swiftness") that makes an appearance in an earlier draft of the poem.[12] Rosenberg's speaker is caught in the "same old" inertia of a waking landscape. "You might object to the second line as vague", the poet wrote to his mentor, Edward Marsh, "but that was the best way I could express the sense of dawn."[13] Twenty years later, prefacing Rosenberg's *Collected Poems*, Siegfried

Sassoon remarked quite simply on the above lines: “Sensuous frontline existence is there.”[14]

In *The Great War and Modern Memory*, Paul Fussell considers the moment Rosenberg tries to capture “an emblem of the political essence of the war itself”.[15] It saw a hustle of activities – reconnaissance, repairing wires, retrieving the dead and wounded from the battlefield – that light would render too dangerous. But sunrise (and sunset) was also the moment of stand-to, when motionless soldiers on both sides peered into no man’s land in case the adversary would take advantage of the lifting fog and semi-darkness to carry out an attack. The day in between they mostly lived a slow, monotonous existence, made up of commonplace activities such as sleeping, letter writing, and sentry duty. Battle, in such an environment, was exceptional. Where the sudden and rapid temporality of combat is often taken as defining conflict, in the popular press especially, the notion of a drawn-out and cyclical temporality Rosenberg articulates in his poem bears on the First World War more widely, characterizing both front and home-front experiences: *slow*, for fighting was locked in a stalemate, in which winning meant wearing the enemy out; and *cyclical*, with combatants rotating between front, support, and reserve lines in a back-and-forth movement we can discern in the war at large, as over a period of four years armies won and lost little terrain at great cost. The conclusion of the war was, in fact, so endlessly protracted that Bertrand Russell suggested contemporaries begin to think of war as a permanent institution in society.

Such slowness registered on the bodies that populated the war world. Consider again Rosenberg’s and Sassoon’s choice of words to explain the poem: “sense”, “sensuous”. If the opening lines to “Break of Day in the Trenches” evoke something of the measured pace and repetitiveness of trench existence, then it is, importantly, a slowness that can be perceived in a visceral way. In a famous 1904 thought experiment the French philosopher Henri Bergson stated that in waiting for sugar to dissolve in water time becomes “no longer something thought; it is something lived”.[16] Boredom was so pervasive in the Great War that for many it did make time a lived experience; temporality weighed heavily on the bodies of soldiers and officers who tried to combat it by smoking, playing cards, or reading tattered magazines.[17] “It is 10 o’clock and time to do something or other”, C.K. Scott Moncrieff, later renowned for his Proust translations, wrote of his unmet expectations of the battlefield, “not that there is anything to do, except look out for a German, and duck your head when you hear a bullet coming.”[18] While civilians did no such ducking, their war hardly moved any faster. For civil servants, it meant paperwork (even soldiering, according to the Futurist painter C.R.W. Nevinson, resembled a kind of office work). For family members, depending on who they were and where they lived, it meant hours of queuing for food, waiting for news and letters, and evacuating for air raids that targeted metropolitan areas like Paris and London. This is in no way to do injustice to the precariousness of their historical moment: even a slow day at the front, or an air-raid alarm in London, carried the threat of danger. But it is to claim that intense, epiphanic moments in the Great War were the exception rather than the rule.

Inactivity came at a cost to the life of the mind, too. Martin Heidegger characterises boredom as "a confrontation with time" that yields a feeling of "being left empty".[19] An anonymous French soldier put his finger on this sense of listlessness as he registered feeling "half-asleep, unconscious, without order, unseeing and unthinking".[20] His words convey aptly what Sianne Ngai means by the hidden potential of 'ugly feelings' to detect 'suspended' or 'obstructed' agencies.[21] Boredom was and to a large extent remains the exemplary experience of those who find themselves on the sidelines of history: not only, as earlier scholarship has shown, women excluded from the public sphere of the early twentieth century and colonial civil servants dismissed to the periphery, but also combatants who were no longer in charge of their own actions and civilians stuck at home.[22] At the same time, the low affect redeems itself in that it shapes aesthetic expression; it is political insofar as it carries within it "an affirmative narrative force".[23] One critic even goes so far as to suggest that the felt need to overcome boredom is at the foundation of the literary enterprise itself: "writing resists boredom, constituting itself by that resistance".[24] In this historical turn to creative expression as a way to safeguard against the loss of self in a protracted conflict – against "unseeing and unthinking" – we can locate some of modernism's most compelling experiments with print culture. The encounter with boredom Rosenberg, Owen, Russell, Scott Moncrieff, and Nevinson discerned between 1914 and 1918 constitutes not only a site for diagnosing socio-political powerlessness, but for rethinking the emergence of modernism from within its affective historical context. For the German critic Siegfried Kracauer, in 1924, boredom was "a necessary precondition for the possibility of generating the authentically new (rather than the old dressed up as the new)".[25]

Printing at the front

Monday, 31 July 1914, Deauville, Normandy:

> A little before midnight I left Deauville
> In Rouveyre's little car
>
> Counting the chauffeur we were three
>
> We said farewell to a whole era
> Furious giants were rising over Europe [...]
>
> And when after spending the afternoon
> Near Fountainebleau
> We arrived in Paris
> Just as they were posting the draft
> We realized my friend and I
> That the little car had driven us into a New era

And although we were both already mature men
We had just been born[26]

Apollinaire's poem was lifted from life. Days before the outbreak of the First World War the poet was on assignment in Deauville, reporting on the summer season for the literary journal *Comœdia*. Upon hearing the announcement of imminent mobilisation, he returned to Paris by car with the painter André Rouveyre. "La Petite Auto" is an impression of that trip. Suggestive of what Enda Duffy calls the "adrenaline aesthetics" of the early twentieth century, the speeding car and the speaker's excitement are replicated in the avant-garde form of the poem, which tosses words and letters across the page.[27] The trip Apollinaire and Rouveyre made symbolises as well the visceral immediacy of their historical moment. The unfolding of the July Crisis of 1914 had been unprecedentedly swift, "unfathomable", as one critic believes, "to anyone who had lived before the age of electronic communication".[28]

Emerging from the First World War is a different Apollinaire. The composition history of "La Petite Auto" indicates that it was not drafted en route to Paris (as it alleges), but in Nice in October, when its author was waiting to enlist in the French forces. After the shock and disruption it had caused in summer, stopping the Parisian avant-garde in its tracks (*Comœdia*, too, folded), the war soon created its own prosaic realities and routines, settling, as it was bound to do, into much slower rhythms. In a lengthy letter – his longest – from the front dated October 1915, Apollinaire described these rhythms as they manifested themselves on an average day in the trenches:

> This, then, is our life: we get up when we want, at least if there's no shelling, and the call "to your positions" finds us quickly ready because we sleep fully dressed. Me, I get up at 7 a.m. only when I'm on duty. When it's my turn, I distribute water, if not, then I have breakfast immediately, black coffee and bread with a cold meal, either Gruyère cheese or marmalade. Then I undress up to the belt and wash myself in a bowl that I have had since the start of the war, an enamelled zinc bowl which is attached to the munitions cart when we move and which has already seen it all. […] After that, I wash my head and rinse it, then my face, then I wash everything that belongs to you, as my wife. Then I dress, then we shell or we read until it's time for supper, then we wait for letters.[29]

While Apollinaire, as patriot and vanguard poet, would never publicly admit to being bored by war ("La Petite Auto" shows the extent to which he felt reborn), the mere length and detailed descriptions of his letter tell a different story.[30] Written in the present tense, its account of the soldier's morning lists the practised movements – "then … then … then … then" – that shaped everyday life at the front (the letter even sports a sketch of the poet's bed, which he goes on to describe at great length). These habits, Annette Becker points out, defy "the long time of the war, which one fragments into the short temporalities of different activities".[31] In two later letters dated February 1916, Apollinaire wrote to his fiancée Madeleine Pagès of the impact of such monotonous living on the mind: "I am today in a state of great melancholia. My watch reads quarter to five, white and cold hour" and "It appears to me that I have dragged my feet through the mud of the great roads for an indefinite time. I become an automaton without veritable thought."[32] The military slang term *cafard*, a soldier suffering from the affliction Apollinaire records, emerged against the backdrop of these individual war experiences.[33]

Apollinaire's war letters implicitly point to the central crisis of a seventeenth-century philosophical tract that travelled far and wide among the educated in the French armed forces: Blaise Pascal's *Pensées*. In part, *Pensées* is a reflection on the need for "divertissement", a sustained effort to distract oneself from the finitude of the human condition. Even "a king without diversion", Pascal suggested provocatively, "is a man full of wretchedness".[34] To divert themselves over long days and years, combatants would turn to letter writing (with family members and *marraines*), reading, masturbation and prostitution, alcohol, and the making of *bagues* and other front souvenirs from shrapnel. It is within this same affective context and as a kind of a mental protective device that the establishment of small trench presses becomes legible as a historical phenomenon. If the conflict had remained the fast-paced war of movement Apollinaire anticipated in "La Petite Auto", then there would have been neither time nor place for prolonged creative expression. Trench presses relied on the slowness of a drawn-out conflict: the stalemate effected after the Battle of the Marne provided an opportunity (and, to be sure, a need) for diversion.[35] Many army periodicals were explicit about this connection between monotonous life and emotional survival, even in their choice of title. The journal of the French prisoners of war at Tauberbischofsheim, for instance, was called *Anti-Cafard*, and each time the war added a year to its record, so did the *Bulletin des Écrivains de 1914*, a magazine sent free-of-charge to all mobilised writers (and to which Apollinaire contributed twice). By the time the magazine folded, after the Armistice, it was known as *Bulletin des Écrivains de 1914–1915–1916–1917–1918–1919.*

These circumstances account for Apollinaire's own turn to the trench press. In June 1915 the poet was stationed in a forest in Champagne, a relatively quiet pocket of the front, where he helped produce the short-lived journal of his battery unit, *Tranchman' Echo*. Later that summer, he would move east towards much livelier fighting, and never found the necessary "leisure to resume the publication of our gazette".[36] Modelled after the *Gazette des Ardennes*, a propaganda magazine the Germans circulated in occupied territories, *Tranchman' Echo* comprised only a handful of pages and was written in a "slightly soldier-like style".[37] Moreover, Apollinaire used the press of his unit to print, together with his soldier-companions Bodard and René Berthier, a volume of poems, *Case d'armons*. The volume contained a small number of *calligrammes* about his war experiences and impressions written in soldier's slang, which Apollinaire wanted to share with the home front (even printing subscription notices before being reminded that commercial activity was prohibited in the French army). The Surrealist Paul Eluard, too, printed 17 copies of *Le Devoir* on a hospital printer at the Somme. Not the printing of a slim volume of verse, but the modernist character of the writing it contained is exceptional in this respect – especially then, especially there.

Publishing journals and poetry volumes in the field was largely possible because of a particular method of printing. "Great Importance of Copying Gelatine", an anonymous piece in the *Mercure de France* of 1 October 1915, alludes to this method and assumes the home front reader's familiarity with it. It is likely that Apollinaire is its author.[38] "Every Frenchman has in his heart a journalist who wakes up from time to time", it begins:

> One of the curious phenomena to which this war has given birth is the publication by front soldiers of journals destined for front soldiers. One designates these ephemeral and sporadic gazettes under the name of trench journals. In reality, they are not edited and printed in the trenches. They see the light of day behind the frontlines (*à l'arrière*), in the villages or the towns where the combatants go on rest. […] In fairness, their general look is not very agreeable to the eye, the printing too often being faulty. […] Trench journals, apart from two or three exceptions, are printed with copying gelatine (*pâte à copier*). We will not describe this procedure, which is well known and which has long been applied in the administration of regiments, companies, batteries, troops. […] Despite all of this, it cannot be predicted that copying gelatine will see its significance grow and maintain after the war. Once the literati are returned to their homes and the typographers to their workshops, the case and linotype will reassert their rights.[39]

Mentioned here only in passing, seeing a trench journal to print was in fact a slow and labour-intensive process. The first two issues of *Tranchman' Echo*, as was the case for many other journals in the French army, were handwritten and existed only in individual numbers that were passed around in the battery, moving down the ranks from officers to gunners to drivers. However, because both documents were lost in unfortunate incidents (one accidentally sent away through the post, one dropped in boiling soup), it was decided to print the next issue of *Tranchman' Echo* on a duplicator bought in a neighbouring village. The procedure went as follows: a master copy was pressed onto a metal tray filled with a copying paste (*pâte à copier*), and the imprint it left in the gelatine was transferred to separate sheets. Only a few dozen copies, often of poor quality, could be produced in this way: of the third issue of *Tranchman'* they printed eight or 12, of *Case d'armons* 25. The latter was further annotated, corrected by hand, individually dedicated, and wrapped for support in pages from the *Bulletin des Armées de la République*, the official gazette of the French army.

My point is that in this slow, careful manual work is located an attempt to authenticate one's humanity not in spite of increasingly mechanised production methods but, rather, in spite of the war's equally modern powers to turn its combatants, as Apollinaire knew, into "automat[a] without veritable thought".[40] Consider this example: as the final issue of *Tranchman' Echo* was unfinished when it went to print (the unit had to move unexpectedly along the lines), the printer stamped the message "particularly targeted by censorship" across its empty columns and blank pages. Clearly a playful gesture, the stamp was also a means to re-take charge of the moment, both in the face of decisions (the move) and regulations (censorship) imposed from higher up. "I was never bored in the war", Apollinaire begins his short account of *Tranchman' Echo*'s origins for the *Mercure de France*, perhaps thinking back of this final issue – a patriotic sentiment, then, clearly a lie (his private correspondence tells otherwise), but also a passing reference to the historical importance of the trench press for shielding men precisely against feelings of *cafard*, the fear of being lived.[41] Trench magazines, in this reading, functioned as a strategy for emotional survival. Moreover, the printing and binding process also offered much-needed relief from the numbing realities of warfare. The makeshift editorial office of the *Wipers Times*, for instance, hidden in a casemate in Ypres, did not only exhibit a printing press, but also had a piano and gramophone. As German shells targeted the Salient, the editors, trying to block out the noise (and their fear), would "turn our whole outfit on together".[42] For Apollinaire, the editing and printing process similarly became a site for the consolidation of affective bonds – the war's "sudden friendships", as a fellow writer named them.[43] René Berthier, whom the poet had met in the barrack dormitory in Nîmes, followed Apollinaire to the front and printed *Case d'armons.*[44]

Because of the non-reifying way in which they were produced, *Tranchman' Echo* and *Case d'armons* remain unique documents – "ephemeral", "sporadic", "faulty", "not very agreeable to the eye" are the words used in the *Mercure de France*. They are tangible archives of life in a particular corner of the frontlines in 1915. With this notion of the archive in mind, Apollinaire sent both documents to the newly established collection of war ephemera at the Bibliothèque nationale de France. In a note, the poet explained that *Tranchman'* was "but interesting as a curiosity, and it is only the fact that it was printed at the front that recommends *Tranchman' Echo* to your indulgence".[45] This observation travels some distance towards explaining why the anecdotal essay in the *Mercure de France* should be titled "Great Importance of Copying Gelatine". While a slow and labour-intensive device, a polycopier is at the same time portable; it can be carried near the fighting (often, near the places of much idleness in a slow war) and thus enabled soldiers to render in print any expression with a certain immediacy. In this way, trench journalists can be viewed as challenging the dreary, numbing pace of the First World War by laying renewed claim to the command of time itself. The first (now lost) issue of *Tranchman' Echo*, for instance, contained "reflections without acrimony on the duration of the war" (*without* acrimony best understood as *with* criticism).[46] The frequency with which Apollinaire's battery journal surfaced – by its own admission: "at each drunken spell of the editor" – supports this point: such sporadic appearances defied not only the regularity of the home press but of the army's scripted, regimented existence.[47] The journal of the No. 3 Canadian Field Ambulance was called *Now and Then*, quite simply because it appeared "now and then, as circumstances permit".[48] Much of the affective potential of the periodical as a genre is premised on this ability to surprise, to divert from the familiar and the expected (that is, from what is *boring*) in a shock-inducing way. In his wartime manifesto, Apollinaire wrote along very similar lines that "[s]urprise is the great new *ressort*. It is by surprise, by the important place it leaves for surprise, that the New Spirit may be recognised."[49]

Apollinaire's slow print in the trenches of the First World War is, then, a political project in the sense that its intricacies of printing and binding enabled the '*cafard*' to restore an individual capacity, a sense of self and agency, including some control over the passing of time. In a short anecdote published in the *Mercure de France* in October 1917, Apollinaire complained of the pressures of modern living, which produced very similar failings:

> Today everyone complains of not having enough time. Authors write their books with haste, before having been able to think about them. [...] For excuse, they say that they don't have the time, that we live in the century of speed, that doing well is not essential: doing it quickly, that's what counts. Speed and progress are the fads of the day [...]. How to rectify this? We only need a bit more slowness; we should leave speed where it belongs, that is, in locomotion. The idea of speed has come to be confused with the idea of progress. [...] All of this also applies to the war, which has collided with all our received notions of contemporary speed. It will be worth it if we kept calling it the war of three months in order to preserve the memory of a mistake due above all to the way in which we have confused speed with progress.[50]

Protracted warfare rendered meaningless contemporary visions of progress and speed. If the trench press as a phenomenon only came into view because of such historical slowness, at least it turned it towards restorative work. Apollinaire's anecdote was joined on the same page of the *Mercure* by his account of the *Gazette Cormon-Collin-Flameng*, the magazine by and for mobilised students of Paris's School of Fine Arts. Another telling example of the poet's initiative to publicise trench journalism at home, the anecdote is attentive to the impromptu material form of the trench journal as well as its content (a poem by "an anonymous *Cafard*", for example, about a soldier on a train who "trembled" instead of "laughed".)[51] By mere juxtaposition the two pieces in the *Mercure de France* may be read to suggest affinity: in this respect, the "*petite revue polycopiée*" of the Parisian students in the trenches answers back to some of the typically modern anxieties the preceding anecdote raises.[52]

Printing at Hogarth House

Tuesday, 9 October 1917, Richmond, London:

> We had a horrid shock. L. came in so unreasonably cheerful that I guessed a disaster. He has been called up. Though rather dashed for 20 minutes, my spirit mounted to a certainty that, save the nuisance, we have nothing to fear. But the nuisance – waiting a week, examination at 8.30 at Kingston – visits to Craig & Wright for certificates – is considerable. It was piteous to see him shivering, so that we lit his gas fire, & only by degrees became more or less where we were in spirits; & still, if one could wake to find it untrue, it would be a mercy.
>
> We took a proof of the first page of K.M.'s story, The Prelude. It looks very nice, set solid in the new type. Masses of bookbinding equipment from Emma Vaughan arrived this morning.[53]

This entry in Virginia Woolf's wartime diary was jotted down precisely a week before the publication of Apollinaire's critique of the "war of three months" in the *Mercure de France.* It gives us an initial notion of what the affective experience of that war was like on the home front across the Channel, as it records the arrival of Leonard Woolf's second call-up, from which he had already been exempted in the previous year. The couple's affective reaction to the news "does a great deal of diagnostic and critical work".[54] Instantly, the Woolfs are struck by a "horrid shock" that manifests itself onto their trembling bodies. The letter's quality to send shivers down the spine lies in its implicit acknowledgement of individual powerlessness. By degree, a less intense, more prosaic sensation emerges, which Woolf once famously described as "this cotton wool, this non-being".[55] Tedious, indeed, is the envisioned encounter with the war's bureaucracies and its culture of paper: appearing in front of the Army Medical Board at the Kingston Barracks, visits to two nerve specialists for letters in support of Leonard Woolf's appeal, and arranging certificates. In *Experience Without Quality,* Elizabeth Goodstein differentiates between the boredom of the factory worker and that of the bourgeoisie, a distinction that applies here as well. For, while both feelings were triggered by warfare, it deserves underlining from the outset that Apollinaire's *cafard,* in the trenches among the *poilus,* is fundamentally different from the "nuisance" experienced by the Woolfs, a well-off pair who enjoyed the comforts of being able to live without a steady income and the related luxury of conscientious objection.

From this initial shock, Woolf's account moves on to the other news of the day: the printing of the first page of Katherine Mansfield's *Prelude* at their newly established Hogarth Press. Mansfield's was, at the time, only the second volume they were printing in their new venture. The idea to establish a small press had famously originated at Virginia Woolf's birthday tea in 1915: "we decided three things: in the first place to take Hogarth, if we can get it; in the second, to buy a Printing press; in the third to buy a Bull dog, probably called John".[56] Short on money at the time, she considered selling her manuscripts of Thackeray's *Vanity Fair*, a family inheritance.[57] Only in early 1917 did the Woolfs enquire about signing up for a printing workshop, from which the couple – "two middle-aged, middle-class persons", as Leonard Woolf remembered the moment in his memoir – were rejected.[58] Without any formal training and after "staring through the window [...] like two hungry children", they had entered a printing shop on Farrington Street on 23 March, and emerged the owners of a small handpress, "The Eclips".[59] The press, broken in two, was delivered during the next month and set up in the drawing room at Hogarth House, while the newspapers recounted the losses and gains of the Battle of Arras. The Woolfs quickly began teaching themselves, kneeling on the drawing-room floor with five pounds of type. The first business was to sort out the type, "a work of ages", as Woolf recalled, slowed down even more when their housekeeper dropped half a case of type on the floor.[60] Virginia Woolf would typically set the type (the reason for Leonard Woolf's exemption from the army was a tremor that made him unfit to carry a rifle as well as set type), and bind and cover the books, for which she used the "masses" of equipment Emma Vaughan sent her (after it had been rejected by the prisoner of war camp on the Isle of Man).[61] Her husband handled the machine and the business side of the venture. Their first publication was a pamphlet announcing the establishment of the Hogarth Press, and while its inking was uneven, its spacing inconsistent, and its punctuation in places lacking, the excitement it gave was tremendous: "We get so absorbed we can't stop; I see that real printing will devour one's entire life", Virginia Woolf noted, calling the process in a letter to Margaret Llewelyn Davies "exciting, soothing, ennobling and satisfying".[62]

Perhaps because of its slow, makeshift character – the false start, the apprenticeship, the flawed product, the naming after a peripheral space – Woolf's work at the Hogarth Press has, as Ursula McTaggart points out, "never [been] coded as political action".[63] Some contemporaries must have assumed the press was a way for the couple to circumvent the publication process. Still, as Lisa Otty remarks on the phenomenon of modernist small presses:

> There were far easier ways to get work published: compositing, distributing type and pulling the press are all physically demanding and time-consuming jobs, on top of which assembling the signatures, arranging binding and distribution had to take place. Most small press owners found themselves devoting copious amounts of time and energy to their operation [...]. Had they simply wanted to put texts into circulation, they surely would have taken the same path as many of their contemporaries and simply employed trade printers, an easier and more cost effective option. For, while it is true that small presses were responding to a fashionable demand for first and limited editions, they were certainly not money-making ventures.[64]

What, then, had the Woolfs been after? The most ready answer that presents itself is an escape from the dictates of the capitalist marketplace that had little taste for modernism; they had wanted the Hogarth Press to be an outlet "for all our friends stories [sic]", which may otherwise not have made it into print.[65] While this may be true perhaps of the 1920s and 1930s, Woolf was not necessarily an experimental modernist in 1917 and experienced no trouble publishing her early work. Moreover, Leonard Woolf recalls in his memoir that "our object was to keep it small", a hobby, then, rather than a legitimate, profit-making business.[66] In line with Apollinaire's contemporaneous experiments in battlefield printing, by contrast, I suggest that the Hogarth Press meant to provide a moment of relief and distraction.[67] It seems that, initially, printing in the drawing room at Hogarth House provided in part a way of coping with a mostly prosaic, sometimes shocking wartime reality. This move from tedium to creativity had been long in the making for the author: from a young age, Woolf would "beguile the dullness [of] innumerable winter walks" by making up stories.[68] "My contribution to the war is the sacrifice of pleasure", the author wrote much later, in May 1940, of what people in the streets were calling 'The Bore War', "I am bored: bored".[69]

Woolf's diary excerpt above suggests as much in its juxtaposition of the shock of the call-up letter with the printing of the first page of Mansfield's *Prelude* and the arrival of the binding equipment, implying that the author's mind is leaping from war to a more life-affirming fixation. The curiously dual affective sensation Woolf records, emphasising both the tediousness and precariousness of life in wartime, also informed her experiences of night-time air raids (an event governed by the same temporality of uncertainty as the unopened letter) which forced the couple, with practised movements, out of their beds and into the cellar. Both civilian experiences – the call-up, the evacuation – register a loss of agency and individuality. War, for Woolf, meant "bodies of human beings" acting "in concert".[70] In light of this reality, work for the Hogarth Press was never uncomplicatedly about the publication of their and their friends' writing, but about the affective process of book production itself: sorting type, inking, printing, binding, covering, with coloured paper and "scissors and paste".[71] The political work of such manual labour – Elizabeth Miller reminds us that "the process of production [is] as politically significant as the product" – lies in its attempt to reassert the labourer's humanity and productiveness (as woman, as conscientious objector), not solely against the machine-like forces of the capitalist marketplace, as others have claimed, but against the equally machine-like forces of war.[72] When a day after their printing press had been delivered to Hogarth House Woolf first met Aldous Huxley, she warned him, upon learning that he had been rejected for military service and had been put in a government office, "of what might happen to his soul".[73] Huxley had replied that there was no need to worry; he spent most of his workday in the administration secretly translating French poetry.

The production of Katherine Mansfield's *Prelude*, the second (and first commissioned) text of the Hogarth Press, is a case in point. The couple was still learning their new craft – Leonard Woolf called it not the business but the "*art* of printing" – and the process was, therefore, incredibly slow.[74] Accepted in July 1917, the Woolfs started printing in November, first 18 pages on their handpress and later on a treadle machine at the Prompt Press in Richmond. The final pages were only printed in July of the next year, and the following week was consumed by "the fury of folding & stapling" 300 copies, of which they would sell 257.[75] The books were sent out on 11 July 1918. However, their slow progress had been interrupted in early December 1917, when a telephone call (another symbol for the temporality of uncertainty, of suspense) informed them of the battlefield death of Leonard Woolf's brother, Cecil. The same shell that killed Cecil had wounded Philip Woolf, the family's youngest. In March, the two surviving brothers and Virginia Woolf began printing a little-known, slim volume of Cecil Woolf's poems from Cambridge, which would become the Hogarth Press's actual second publication. They produced only five copies.[76] None of these were sold at the time, confirming the suspicion that it was not really about the poems (and never about capitalist interests), but about the intimate process of printing. As J. H. Willis Jr. argues, producing the booklet was "a family gesture" and "a therapy for Philip", who helped set the type and wrap the books in white paper.[77] In this sense, Cecil Woolf's *Poems* matters as a personal archive. In a parallel gesture of commemoration, John Middleton Murry's Heron Press appears to have been named after Katherine Mansfield's brother, Leslie Beauchamp, whose middle name was Heron and who was killed in fighting at Ploegsteert Wood in 1915.

Gradually, the press began, as Woolf put it, "to outgrow its parents".[78] For McTaggart, that pivotal moment was 1920, when the press moved from small-scale printing to commercial publishing; it could just as easily have been 1919, when Leonard Woolf noted how the success of *Kew Gardens* led them, indirectly and reluctantly, to turn the Hogarth Press into a commercial printing business and when Virginia Woolf, in turn, remarked for the first time that the work was "a little overwhelming" and that she thought of setting up an office and hiring a secretary.[79] It could have been 1921, when the Woolfs replaced their handpress with a second-hand Minerva platen machine, or 1922, when the business published its first novel, *Jacob's Room*, and considered merging with Heinemann. For Willis Jr., the sea change came in March 1924, when the enterprise moved to Tavistock Square in London. Whichever it was, the point is that these moments all came after the First World War, and that, taken separately and together, they constitute a move away from what began, in mid-war, as an intimate, two-person venture in the drawing room at Hogarth House. By 1938, now an established voice, Virginia Woolf recalled, melancholically, "Press worth £10,000, & all this sprung from that type on the drawing room table at Hogarth House 20 years ago".[80]

Epilogue

"The horror of the years 1914 to 1918", Leonard Woolf recalled in *Beginning Again*, "was that nothing seemed to happen, month after month and year after year, except the pitiless, useless slaughter in France."[81] These words return, almost verbatim, in what is often regarded as Virginia Woolf's first modernist story. "The Mark on the Wall" sets its scene in a domestic environment, where a bored protagonist notices the eponymous stain on the wall of her living room. But rather than figure out what it might be or where it might come from, her thinking refuses to come into focus ("let me just count over a few of the things lost in one lifetime") until her husband exclaims: "*Nothing ever happens* [...] Curse this war; God damn this war! ... All the same, I don't see why we should have a snail on our wall."[82] This epiphanic moment, the intrusion of the war in the narrative, has an unsettling effect for both protagonist and reader. Wartime affective life was shaped by sensations of boredom – the symbol of the snail resembles Apollinaire's drawing of his dug-out bed in this respect – as well as by sudden shock-like reminders of the precariousness of human life. The intimate publication process of "Mark" in *Two Stories* – type-set, hand-printed in the Woolfs' drawing room, carefully wrapped in "rather unusual, gay Japanese" cover paper – can be read to offer a moment of release from the kind of wartime realities captured in the story. So much is apparent from the faulty, tangible book itself: for William Rothenstein, it exuded "a more human quality".[83]

Yet, moving beyond these singular cases, a consideration of the production process of a book like *Two Stories, Case d'armons*, or Cecil Woolf's *Poems* compels us to think anew the relationship between temporality, affect, and modernism. First, such an approach will have to attest to the significance of low affects for our understanding of the emergence of modernism within its historical context. To do precisely this is to follow a trajectory Virginia Woolf outlined in a 1926 essay, in which she called for "a new hierarchy of the passions", noting the need for a "new language [since] English [...] has no word for the shiver and the headache".[84] The shiver, then, brings into view – "diagnoses", for Sianne Ngai – a less confident, much messier reality, one in which modernity as a set of ideas and values is found to be failing. The First World War, in many ways, was such a moment. Second, a turn to low affects helpfully complicates one of the central tenets in modern writing: its fixation with temporality. There are many more temporalities at work in modernist texts and contexts than its now-clichéd logic of speed and immediacy proposes. The slow, thick temporality implied in the feeling of boredom – in a "war of three months", as Apollinaire termed it, preserving "the memory of a mistake" – is only one such example. These alternative temporal orders figure not only in writing, but also in the creation processes and even in the emergence of the phenomenon itself: if, as Enda Duffy argues, speed had become "the very narrative heft of much modernist artistic production", then Apollinaire's and Woolf's wartime experiments with printing throw into sharp relief the much slower pace with which experimental art came to the scene.[85]

Notes

1 For various kindnesses, I want to thank Annette Becker, Laurence Campa, and Sarah Posman.

2 Apollinaire, *Œuvres poétiques*, 674.

3 Woolf, *The Diary of Virginia Woolf, Vol. I*, 85.
4 Ngai, *Ugly Feelings*, 27.
5 Woolf, *The Diary of Virginia Woolf, Vol. I*, 116.
6 Woolf, *The Diary of Virginia Woolf, Vol. I*, 85.
7 Auerbach, "Imperial Boredom", 302.
8 Willis, *Leonard and Virginia Woolf as Publishers*, 41.
9 Southworth, *Leonard and Virginia Woolf, the Hogarth Press and the Networks of Modernism*, is representative of this approach.
10 Share and Wiman, *The Open Door*, 75.
11 Owen, *The Complete Poems and Fragments*, 174.
12 Share and Wiman, *The Open Door*, 75.
13 Noakes, *Isaac Rosenberg*, 308.
14 Sassoon, "Foreword", xi.
15 Fussell, *The Great War and Modern Memory*, 52.
16 Bergson, *Creative Evolution*, 1.
17 For instance, Maeland and Brunstad, *Enduring Military Boredom*. On the importance of morale, see Fuller, *Troop Morale and Popular Culture in the British and Dominion Armies*.
18 Scott Moncrieff, *Memories and Letters*, 55.
19 Pease, *Modernism, Feminism and the Culture of Boredom*, 5.
20 Audoin-Rouzeau, *Men at War*, 41.
21 Ngai, *Ugly Feelings*, 1, 3.
22 Pease, *Modernism, Feminism and the Culture of Boredom*; Auerbach, "Imperial Boredom"; Majumdar, *Prose of the World*.
23 Majumdar, *Prose of the World*, 12.
24 Spacks, *Boredom*, 1.
25 Kracauer, *The Mass Ornament*, 334.
26 Apollinaire, *Calligrammes*, 105–111. The first line of the poem actually reads "31 August 1914", which has puzzled critics for over a century now (Apollinaire, *Œuvres poétiques,* 1085–1086).
27 Duffy, *The Speed Handbook*, 9.
28 Kern, *The Culture of Time and Space*, 260.
29 Apollinaire, *Correspondance générale*, II, 885.
30 Interestingly, the slowness of Apollinaire's historical context began to inhabit his language: "la guerre continue, continue"; "je crois que cette semaine va me paraître longue, longue"; "je suis fatigué, fatigué" (Apollinaire, *Correspondance générale*, II, 564, 30, 59). These repetitions appear to occur in those passages that address the war's slow monotonies of trench life, enacting in miniature an issue Vincent Sherry diagnoses in the language of modernism at large: "the representation of temporality [...] dissolves chronology into the sheer plod of one damn thing after another, the merely serial quotidian" (Sherry, *The Great War and the Language of Modernism*, 267).
31 Becker, *La Grande Guerre d'Apollinaire*, 119.
32 Apollinaire, *Correspondance générale*, III, 92, 98.
33 On the contemporary understanding of the affliction, see Huot and Voivenel, *La Psychologie du soldat* and *Le Cafard*.
34 Pascal, *Pascal's Pensées*, 43–44.
35 The editor of *Tacatacteufteuf* motivated the founding of his journal as follows: "Killing time is not an aim. Yet this is our only raison d'être. Killing the long hours of inactivity at the front, killing the *cafard* [...] to escape the daily constraints and distresses that come his way as part of the unending dangers of his life" (Audoin-Rouzeau, *Men at War*, 18).
36 Apollinaire, *Œuvres en prose complètes*, III, 250. Apollinaire recounts the origin story of the magazine in a short anecdote entitled "L'Histoire d'une gazette du front" in the *Mercure de France*.

37 Apollinaire, *Œuvres en prose complètes*, III, 248.

38 A number of reasons lead me to assume Apollinaire or someone close to him authored this piece. First, it mentions *Case d'armons*, an obscure, little-known volume in 1915 (before *Calligrammes* was published) and *Les Imberbes*, an even less-known trench journal to which Apollinaire also contributed (though his poem never appeared). Second, *Échos* was one of Apollinaire's columns in the *Mercure de France*, which he took over from Remy de Gourmont. Third, the piece dates from around the time of the poet's first encounters with the trench press (its publication in October 1915 means that the piece was probably written over the summer).

39 Anon., "Grande Importance de Pâte à Copier", 397–398.

40 Interestingly, Walter Benjamin is both the author of *The Work of Art in the Age of Its Technological Reproducibility* (1936), which describes the impact of these mechanised production methods, as well as one of the earlier theorists of boredom in the modern age. In that latter affective experience, Benjamin discerns a similar space for potential creativity – what other critics term a motive power, a narrative force.

41 Apollinaire, *Œuvres en prose complètes*, III, 247.

42 Roberts, "How it happened", v.

43 Dos Passos, *The Best Times*, 57.

44 "My bed is close to that of a brigadier-poet", Apollinaire wrote of his first encounter with Berthier in Nîmes (Apollinaire, *Œuvres en prose complètes*, III, 21). They would each go on to write about their friendship in Parisian periodicals: Apollinaire about Berthier in the *Mercure de France* and Berthier about Apollinaire in the Parisian vanguard monthly *SIC*.

45 Apollinaire, *Correspondance générale*, II, 765.

46 Apollinaire, *Œuvres en prose complètes*, III, 248.

47 The quote is from the head mast of the remaining copy of *Tranchman' Echo* in the Bibliothèque nationale de France. There is perhaps an opportunity here to read the trench press as reclaiming the "reflective pause" between consecutive issues of a periodical, which, as Mark Turner has argued, became the point of much contention and anxiety in modernity (with the unwelcomed move from quarterlies and monthlies to weeklies and dailies leaving too little time for contemplation). See Turner, "Periodical Time in the Nineteenth Century", 194.

48 See the front-page editorial of the first issue (December 1915). In part, trench publications owed their irregularities to the unpredictable circumstances of war, including censorship, the death of contributors, the destruction or loss of printing material, and paper shortages.

49 Apollinaire, *Œuvres en prose complètes*, II, 948.

50 Apollinaire, *Œuvres en prose complètes*, III, 259–260.

51 Apollinaire, *Œuvres en prose complètes*, III, 262.

52 Apollinaire, *Œuvres en prose complètes*, III, 261.

53 Woolf, *The Diary of Virginia Woolf, Vol. I*, 56–57.

54 Ngai, *Ugly Feelings*, 29.

55 Woolf, *Moments of Being*, 84.

56 Woolf, *The Diary of Virginia Woolf, Vol. I*, 28.

57 Leslie Stephen's first wife, Harriet Thackeray, was the daughter of the famous novelist.

58 Woolf, *Beginning Again*, 233.

59 Woolf, *Beginning Again*, 234.

60 Woolf, *The Diary of Virginia Woolf, Vol. I*, 143.

61 This handiwork shaped Woolf's modernist style. Setting type for Eliot's *Poems* and *The Wasteland* or Hope Mirlees' *Paris: A Poem* added to Woolf's understanding of the modernist page, and it influenced in particular her use of white space in *Jacob's Room*, a book that she began conceiving after reading Maynard Keynes' *The Economic Consequences of the Peace*. See Marcus, "Virginia Woolf and the Hogarth Press", 131.

62 Woolf, *The Letters of Virginia Woolf, Vol. II*, 150–151.
63 McTaggart, "Opening the Door", 64.
64 Otty, "Small Press Modernists", 130.
65 Woolf, *The Letters of Virginia Woolf, Vol. II*, 159.
66 Woolf, *Beginning Again*, 4.
67 Virginia Woolf was prone to mental illness. "It would be good if Virginia had a manual occupation of this kind which, in say the afternoons, would take her mind completely off her work" (Woolf, *Beginning Again*, 233).
68 Woolf, *Moments of Being*, 89. Sara Crangle, too, sees boredom as a "catalyst for creation", both for Woolf herself as well as for the characters in her fiction (Crangle, *Prosaic Desires*, 101). In her incredibly informative *Prosaic Desires*, she investigates the desire to be relieved from boredom as shaping modernist writing.
69 Woolf, *The Diary of Virginia Woolf, Vol. V*, 288; Levenback, *Virginia Woolf and the Great War*, 154.
70 Woolf, *The Diary of Virginia Woolf, Vol. I*, 229.
71 Woolf, *The Letters of Virginia Woolf, Vol. II*, 436.
72 Miller, *Slow Print*, 57.
73 Woolf, *The Letters of Virginia Woolf, Vol. II*, 150.
74 Woolf, *Beginning Again*, 233. This is, in fact, in line with Virginia Woolf's writing process: "When I remember how, owing to her health, Virginia always had to restrict her daily writing to a few hours and often had to give up writing for weeks or even months, how slowly she wrote and how persistently she revised and worked over what she had written before she published it, I am amazed that she had written and published seventeen books before she died" (Woolf, *Downhill All the Way*, 58).
75 Woolf, *The Diary of Virginia Woolf, Vol. I*, 164.
76 The book was reviewed in *The Times Literary Supplement*. The reviewer seemed to appreciate the inexpert writing and private printing: "And in the existence of this frequent practice of poetry by 'amateur' poets lies the source of strength for English poetry now and in the future" (Anon., "Four Young Poets", 40).
77 Willis, *Leonard and Virginia Woolf as Publishers*, 23–24. In her diary, Virginia Woolf writes: "as giving him [Philip Woolf] occupation it would be a good thing to do", admitting at the same time that Cecil Woolf's poems were "not good; they show the Woolf tendency to denunciation, without the vigour of my particular Woolf" (Woolf, *The Diary of Virginia Woolf, Vol. I*, 123–124).
78 Woolf, *The Diary of Virginia Woolf, Vol. II*, 144.
79 Woolf, *The Letters of Virginia Woolf, Vol. II*, 368.
80 Woolf, *The Diary of Virginia Woolf, Vol. V*, 137.
81 Woolf, *Beginning Again*, 197.
82 Woolf, *The Mark on the Wall and Other Short Fiction*, 4, 10.
83 Willis, *Leonard and Virginia Woolf as Publishers*, 18.
84 Woolf, *Selected Essays*, 102.
85 Duffy, *The Speed Handbook*, 10.

Bibliography

Published materials

Apollinaire, Guillaume. *Œuvres poétiques*, edited by Marcel Adéma and Michel Décaudin. Paris: Gallimard/Bibliothèque de la Pléiade, 1956.

Apollinaire, Guillaume. *Calligrammes: Poems of Peace and War (1913–1916)*. Translated by Anne Hyde Greet. Berkeley: University of California Press, 1980.

Apollinaire, Guillaume. *Œuvres en prose complètes*, II and III, edited by Pierre Caizergues and Michel Décaudin. Paris: Gallimard/Bibliothèque de la Pléiade, 1991 and 1993.

Apollinaire, Guillaume. *Correspondance générale*, I–III, edited by Victor Martin-Schmets. Paris: Honoré Champion, 2015.

Anon. "Grande Importance de Pâte à Copier." *Mercure de France* 112, no. 418 (1 October 1915): 397–398.

Anon. "Four Young Poets." *The Times Literary Supplement*, no. 888 (23 January 1919): 40.

Audoin-Rouzeau, Stéphane. *Men at War: National Sentiment and Trench Journalism in France during the First World War*. Oxford: Berg, 1995.

Auerbach, Jeffrey. "Imperial Boredom." *Common Knowledge* 11, no. 2(2005): 283–305.

Becker, Annette. *La Grande Guerre d'Apollinaire: Un Poète combattant*. Paris: Texto, 2014.

Benjamin, Walter. *The Work of Art in the Age of Its Technological Reproducibility, and Other Writings on Media*. Edited by Michael W. Jennings, Brigid Doherty, and Thomas Y. Levin. Cambridge, MA: Harvard University Press, 2008.

Bergson, Henri. *Creative Evolution*. New York: Cosimo, 2005.

Berthier, René. "Espoir en Guillaume Apollinaire." *SIC* 1, no. 11(August 1916): 10.

Campa, Laurence. *Guillaume Apollinaire*. Paris: Gallimard, 2013.

Crangle, Sara. *Prosaic Desires: Modernist Knowledge, Boredom, Laughter, and Anticipation*. Edinburgh: Edinburgh University Press, 2010.

Dos Passos, John. *The Best Times: An Informal Memoir*. London: André Deutsch, 1968.

Duffy, Enda. *The Speed Handbook: Velocity, Pleasure, Modernism*. Durham: Duke University Press, 2008.

Fuller, J.G. *Troop Morale and Popular Culture in the British and Dominion Armies*. Oxford: Clarendon, 1991.

Fussell, Paul. *The Great War and Modern Memory*. Oxford: Oxford University Press, 1977.

Goodstein, Elizabeth. *Experience Without Quality: Boredom and Modernity*. Stanford: Stanford University Press, 2005.

Huot, Louis and Paul Voivenel. *Le Cafard*. Paris: Bernarnd Grasset, 1918.

Huot, Louis and Paul Voivenel. *La Psychologie du soldat*. Paris: La Renaissance du Livre, 1918.

Kern, Stephen. *The Culture of Time and Space 1880–1918*. Cambridge, MA: Harvard University Press, 1983.

Kracauer, Siegfried. *The Mass Ornament: Weimar Essays*. Cambridge, MA: Harvard University Press, 1995.

Levenback, Karen L. *Virginia Woolf and the Great War*. Syracuse: Syracuse University Press, 1999.

Maeland, Bard and Paul Otto Brunstad, *Enduring Military Boredom: From 1750 to the Present*. Basingstoke: Palgrave Macmillan, 2009.

Majumdar, Saikat. *Prose of the World: Modernism and the Banality of Empire*. New York: Columbia University Press, 2013.

Marcus, Laura. "Virginia Woolf and the Hogarth Press." In *Modernism and the Marketplace*, edited by Ian Willison, Warwick Gould, and Warren Chernaik, 124–150. Basingstoke: Macmillan, 1996.

McTaggart, Ursula. "'Opening the Door': The Hogarth Press as Virginia Woolf's Outsiders' Society." *Tulsa Studies in Women's Literature* 29, no. 1(Spring 2010 2010): 63–81.

Miller, Elizabeth C. *Slow Print: Literary Radicalism and Late Victorian Print Culture.* Stanford: Stanford University Press, 2013.
Ngai, Sianne. *Ugly Feelings.* Cambridge, MA: Harvard University Press, 2005.
Noakes, Vivien (ed.). *Isaac Rosenberg.* Oxford: Oxford University Press, 2012.
Otty, Lisa. "Small Press Modernists: Collaboration, Experimentation and the Limited Edition Book." In *The Aesthetics of Matter: Modernism, the Avant-Garde and Material Exchange*, edited by Sarah Posman, Anne Reverseau, David Ayers, Sascha Bru, and Benedikt Hjartarson, 128–141. Berlin: De Gruyter, 2013.
Owen, Wilfred. *The Complete Poems and Fragments*, edited by Jon Stallworthy. London: Chatto & Windus, 2013.
Pascal, Blaise. *Pascal's Pensées.* Translated by W.F. Trotter and with an introduction by T.S. Eliot. London: J.M. Dent, 1931.
Pease, Allison. *Modernism, Feminism and the Culture of Boredom.* Cambridge: Cambridge University Press, 2012.
Roberts, R.J. "How It Happened." In *The Wipers Times.* London: Herbert Jenkins, 1918.
Sassoon, Siegfried. "Foreword." In *The Collected Works of Isaac Rosenberg*, edited by Ian Parsons. London: Chatto & Windus, 1984.
Scott Moncrieff, C.K. *Memories and Letters*, edited by J.M. Scott Moncrieff and L.W. Lunn. London: Chapman & Hall, 1931.
Share, Don and Christian Wiman. *The Open Door: One Hundred Poems, One Hundred Years of "Poetry" Magazine.* Chicago: University of Chicago Press, 2012.
Sherry, Vincent. *The Great War and the Language of Modernism.* New York: Oxford University Press, 2003.
Southworth, Helen (ed.). *Leonard and Virginia Woolf, the Hogarth Press and the Networks of Modernism.* Edinburgh: Edinburgh University Press, 2010.
Spacks, Patricia Meyer. *Boredom: The Literary History of a State of Mind.* Chicago: University of Chicago Press, 1995.
Turner, Mark. "Periodical Time in the Nineteenth Century." *Media History* 8(2002): 183–196.
Willis, J.H. *Leonard and Virginia Woolf as Publishers: The Hogarth Press, 1917–41.* Charlottesville: University Press of Virginia, 1992.
Woolf, Leonard. *Beginning Again: An Autobiography of the Years 1911–1918.* London: The Hogarth Press, 1964.
Woolf, Leonard. *Downhill All the Way: An Autobiography of the Years 1919–1939.* London: The Hogarth Press, 1975.
Woolf, Virginia. *The Diary of Virginia Woolf, Vol. I: 1915–1919*, edited by Anne Olivier Bell. London: The Hogarth Press, 1977.
Woolf, Virginia. *The Diary of Virginia Woolf, Vol. II: 1920–1924*, edited by Anne Oliver Bell. London: The Hogarth Press, 1978.
Woolf, Virginia. *The Letters of Virginia Woolf, Vol. II: 1912–1922*, edited by Nigel Nicholson. London: Chatto & Windus, 1980.
Woolf, Virginia. *The Diary of Virginia Woolf, Vol. V: 1936–1941*, edited by Anne Olivier Bell. London: The Hogarth Press, 1984.
Woolf, Virginia. *The Mark on the Wall and Other Short Fiction*, edited by David Bradshaw. Oxford: Oxford University Press, 2001.
Woolf, Virginia. *Moments of Being: Autobiographical Writings*, edited by Jeanne Schulkind. London: Pimlico, 2002.
Woolf, Virginia. *Selected Essays*, edited by David Bradshaw. Oxford: Oxford University Press, 2008.

Part II
Reorientation and memory

5 "It is at night-time that we notice most of the changes in our life caused by the war"

War-time, Zeppelins, and children's experience of the Great War in London

Assaf Mond[1]

On the night of 8 September 1915, as the second year of the First World War began, the Zeppelins raided London for the fourth time. When the bombs started to fall, 12-year-old Jack Marriage, who lived in 2 Beauchamp Street, was woken by his mother with an alarming warning. As he recalled it in an essay he wrote just a few days later: "On Wednesday night, at quarter to eleven, I was woken up by my mother, who said: 'don't be frightened, the Germans are here.'"[2]

What Jack's mother said hides several deeper meanings, and sheds some light on daily life in London during the First World War. Firstly, it expresses the perception of invasion: the arrival of German soldiers borne in a German war machine to the heart of the empire symbolized a new era of aerial warfare, in which Britain's fleet could no longer protect it.[3] Moreover, his mother's warning – "the Germans are here" – symbolized the war's tangible entrance into the urban sphere, in a new era of 'total war'. The Zeppelins – followed by the German aeroplanes – brought sights of death and destruction to the British home front, changed the city's landscape with their bombs, and added to modern warfare the component of strategic bombardments (now mostly identified with the London Blitz in the Second World War). In that sense, the Germans' arrival was both real and a metaphor for the emergence of the war in London's urban life. "The Germans are here" thus represents a notion of a larger scale: the *war* is here.

While the 'First Blitz' was not as lethal as the infamous Blitz of the Second World War, it was during the first period that the foundations were laid for both military and civic deployments for future aerial raids.[4] During the First World War, airships carried out a total of 51 raids on Britain, in which 6,000 bombs totalling 200 tonnes were dropped over the nation. In those raids, 557 people were killed and 1,358 injured. London was bombed nine times, causing the deaths of 181 Londoners and a further 504 injured. As of the end of 1916, however, the Zeppelins became a lesser threat to British civilians than the German aeroplanes, which caused the death of 857 British civilians – more than the Zeppelins, in roughly the same amount of raids. Those numbers seem negligible compared with the casualties on the battlefields of the

Western Front, but the impression given by the casualty statistics must be contrasted with the psychological damage resulting from months and years of the blackout regime, and fear of the appearance of a 650-foot-long airship in the skies. On the other hand, concomitant with dread of the bombs, there was excitement at the sight of cruising technology. Curiosity often impelled people to go out and gaze at the modern phenomenon, overcoming both the fear and the instructions to take cover.[5]

While the Great War remains one of the most studied events in history, the airships have so far been side-lined in research – and sometimes they merit just a few lines in weighty textbooks which claim to provide a complete picture of the war. One possible explanation may be the relatively low number of casualties they inflicted. Another cause for the small number of historical and cultural studies about the Zeppelin raids may lie in the numerous books written about German airships as part of military history.[6] Possibly, they generated the mistaken impression that it was a worn-out topic. A third possible reason for the paucity of references to Zeppelins in cultural history research is the airship's place in modern collective memory: the 1937 crash of the LZ-129 Hindenburg in New Jersey imprinted a visual memory of the Zeppelin as a lumbering, obsolete machine. Perhaps it led people to forget that from 1915 to 1918 it was an advanced modern means of warfare that constituted a genuine threat – a notion that had only recently received a prominent place in the research about the British home front.[7] Therefore, the Zeppelins should be part of the modern research about the reactions to the outbreak of the First World War[8] and about the cultural history of civilians in the war.[9]

This chapter is part of a wider study about the conceptual and physical entrance of the Great War into the London urban sphere. I examine the Zeppelin experience of Great War London as reflected in diaries written by Londoners during the war, and in the essays by young Jack Marriage and his school friends. The Zeppelin experience, as I show, changed the way London residents perceived the urban sphere from the early days of the war, when the aerial raids were merely an imagined threat, and through the actual attacks by which the Zeppelins brought the war to the city, in the most physical way.

The first part of this chapter presents my argument about the overall significance of the threat of Zeppelin raids, which changed the way time was perceived in London during the war and forced Londoners to rearrange their daily life time-frames. The differences between day and night, for example, were no longer the same once the darkness of night-time was marked as the preferred cover for the raiders, and after a strict blackout regime was enforced to conceal the city from its attackers. At the beginning of the war, the Zeppelins were still a distant threat. However, the new 'state of emergency' forced on Londoners a change of time perception, between 'peace time' and 'war time'. When the air raids became a closer threat, the blackout regime and restrictions on opening hours of pubs, for example, changed the way Londoners defined 'day time' and 'night time' – and enhanced the changing perceptions of time in the city.

In the second part, I discuss the actual raids, and argue that the Zeppelins blurred the boundaries between the distinct worlds of children and of adults in Great War London. During 'peace time', the 'day time' was the time of children's activity, while 'night time' was the time they were supposed to go to bed. However, as I show, the aerial raids changed that separation in the world of children, by turning the 'night time' into a time of danger and threatening their childhood. When the airships raided, children were woken up in the middle of the night, helping their parents to take care of their younger siblings, or walking outside underneath flying weapons of war, witnessing dead bodies and severe injuries. Fortunately, a remarkable source allows us to grasp something of what the air raids meant for children. Educational psychologist Charles William Kimmins conducted during the war studies on children's reactions to air raids, which allows us to chart this early childhood experience of nocturnal 'war time' and the ways it changed what they saw and challenged how they acted. The accounts of those children can enrich our understanding of the Zeppelin experience and war experience of London children during the Great War – as an early example of targeting of civilian populations from the air, which was to become a mass experience in the course of the twentieth century.

So far, very little was written about Kimmins and his studies of London children during the Great War. The 48 essays that I discuss here, which were written by schoolboys from Princeton Street School in Bedford Row, Holborn, have not so far been known to be part of Kimmins' studies, and were rarely cited for themselves. Through those essays, and through Kimmins' books and academic articles, I show how the life of London children underwent changes because of the raids and the active roles they assumed in adult spheres of time and war.

The schoolboys of Princeton School, aged 6–14 – the common definition of the age of childhood at the time – were part of the generation that was born between 1900 and 1910, grew up during the war, and many times lived in a home without a father or an older brother. They were victims of the war in many aspects – being under the threat of the Zeppelin bombs was just one of them – and they grew up to be the adults of the Second World War, which means that they were the embodiment of the way the Great War shaped the perception of total war in the inter-war years and in 1939. However, as Manon Pignot observed, "It was not so much the war itself that had an impact, as a certain sort of experience of the war that crystallised children's reality and left long-lasting traces on the adults to come."[10] Meaning: in dealing with the world of children and the way it was changed because of the war, the emphasis should be on the specific experiences that they were part of during the war – like an aerial raid.

The Zeppelin was both an imagined threat and an actual one. The urban sphere of London was changed when Britain prepared to be raided, and also when the raids began and brought the sights of the front to the home front. This can be seen through the diaries of the Londoners who experienced this physical change in their urban sphere, and perhaps mostly through the rare essays by the city's children, who wrote in length about their perception of the war scene in the middle of the urban sphere. The evidences of those children are not only a unique voice in the history of the Great War, but also an authentic and unbiased one – because they were part of Kimmins' psychological study, and were written under his strict orders. The way these children understood time and space during the raid, therefore, can enrich our understanding of the Zeppelin experience, time perceptions, and the life in Great War London.

Part I

"As if time itself was standing still"

On 1 August 1914, Hallie Eustace Miles and her husband were away from their home in Chandos Street, London, and were enjoying a short holiday at Westgate-on-Sea. "The weather is glorious," she wrote in her diary. "There seems nothing wrong (outwardly) with the world in general, yet the air is full of whispers of coming trouble: and the rumours of War are becoming more and more alarming and more and more persistent." Miles wrote her next diary entry three days later, after Britain declared war on Germany. "Since writing those words, we already seem to be in a new world," she wrote, and explained the decision to cut short their holiday:

> The awful Declaration of War has been made, and everything is plunged into chaos. At first one feels paralysed; as if time itself was standing still. We are at once leaving these quiet scenes, for we feel we want to be in the midst of things in London.

The following entry, accordingly, was written in the British capital:

> We are back in London. In the short time that we have been away everything seems to have changed! It is like an altogether new London. The whole condition of things is so strange to us all; for to have *London* in a state of War is not within the memory of the present generation.[11]

Those few lines depict how the war – even the idea of war – had started affecting perceptions of space and time. Over the span of three days, as Miles wrote, "everything seems to have changed." When the outbreak of a war was no more than a threat, nothing seemed to have changed with the world "outwardly," but as soon as war was declared, she felt "as if time itself was standing still" – and when this paralysing feeling disappeared, it was already "a new world." On 4 August 1914 Britain's part in the war was still merely symbolic – for the British army was yet to reach the front, and the home front had not yet been attacked – but the couple hurried up to London to be "in the midst of things." Once they reached the capital city, they noticed that it was a "new London." The city had not yet come under attack, but recent events were already changing the way Miles felt in it.

The moment of change was described in the diary as a halting of time, which separated between two time-frames: until 4 August 1914 London was in a 'time of peace', but then began an era of "*London* in a state of War," or a spatial 'time of war'. The perceptions of 'war time' and 'peace time' in London were based mostly on Miles' and other Londoners' feelings towards the urban sphere, but also had some physical manifestations. Perhaps the most vivid way in which time was presented in the London sphere was through the Elizabeth Tower and its clock, commonly known as Big Ben.

War was to be declared at midnight, German time, on 3 August, which was 11 p.m. British time. Michael MacDonagh, a reporter for *The Times* who lived on Abbeville Road, wrote in his diary:

> When the ultimatum to Germany was set to expire, we returned in our thousands to Whitehall [...] Then followed the slow and measured strokes of Big Ben proclaiming to London that it was eleven o'clock [...] Perhaps it was but a reaction to the mood we were in, but I thought Big Ben was tolling the hour with an even more solemn note [...] Was he booming out sweet peace and in red slaughter?[12]

Here, the strokes of Big Ben represent a countdown towards the transition from 'peace time' to 'war time' – and it was the absence of these strokes and the disappearance of time from the city soundscape that would symbolize Great War London. "There is one thing you never hear now – the bells of the public clocks telling the passing of the hours," wrote MacDonagh in his diary on 3 January 1915. "Some of the clocks have been stopped going altogether. The others record the time on their faces, but they have ceased to tell it audibly. At night they are darkened. Why? [...] that the Zeps may not know the place or the time when they come to London!"[13] The 'war time', which at its beginning was sensed by Miles as if "time stood still" – a phrase that was also used during the war by prisoners of war[14] – forced the actual stoppage of many public clocks, and the silencing of others.

The clocks did not strike – and the silence was felt by London residents. "I notice the absence of the chimes of Big Ben at the end of each hour at the office," wrote on 9 October 1914 the young civil servant who published his diary under the pseudonym Victor Smith,[15] while Miles described the "intensely solemn" New Year's Eve at the end of 1915: "no clashing bells were to be heard; even Big Ben was silent."[16] MacDonagh's diary entry for 21 May 1916 notes that the silenced clock was even humorously personified as the defender of Whitehall against the aerial raiders. "Ware Zeppelins," he wrote. "It would not do for Big Ben to betray the House of Parliament to the marauders."[17]

On the same day, 21 May 1916, the absence of clock chimes was not the only time-related issue that perturbed MacDonagh – and once again the Zeppelins were blamed for changing the time-frames on the home front. "This is the first day of 'summer-time' as by law established," he wrote. "We put our clocks forward an hour after midnight last night so that we might have the additional hour of daylight in the evenings."[18] This first implementation of Daylight Saving Time was mainly justified as a measure to assist the wartime economy – since longer daylight meant that less coal was needed for artificial lighting, and also gave the working classes more opportunities for open-air recreation – but the delaying of the darkened hours also increased the part of the day that was safe from the airship, which only raided at night.[19] "In most ways it is really delightful," Miles wrote about it a month later, and reflected on the forced change in the frames of time at the home front. "I shall never forget the strange and solemn moment that it was, when I put all our clocks and watches on, one hour. It seems as if I was interfering with Nature, and pushing on the Sun and Moon and Stars."[20]

Those changes, and especially the silencing of the clocks, were perceived in the urban sphere to be a result of the war, and therefore part of the temporary 'war time' that would eventually change once again. "When I cannot sleep, I try to picture to myself the distant time when the Days of Peace will be with us again," wrote Miles on 1 January 1916. "In my imagination [...] I hear the bells daring to peal and clash in the old sonorous way, and Big Ben striking the hours again."[21]

The soundscape of Great War London was changed because of the transition from 'peace time' to 'war time', the result being that while time may not have stood still – it stood silent. That silence was attributed to the 'war time', and was cheerfully broken – as Miles had hoped – after the Armistice. And so, on 11 November 1918, Michael MacDonagh hurried once again to face the clock tower. "I had heard Big Ben proclaiming war," he recalled his presence in Whitehall on the night of 3 August 1914.

> I was now to hear him welcoming peace [...] when the hands of the dials pointed to XII, Big Ben struck the hour, booming it in his deep and solemn tones, so old and so familiar. It was most dramatic moment. The crowd that had assembled in Parliament Square stood silent and still until the last stroke of the clock, when they burst into shouts of exultation.[22]

The 'war time' came to an end, and 'peace time' was once again heard in London.

"London has been blacked out"

"During the autumn and winter of 1914 we began to suffer from a curtailed supply of fuel and light," wrote Constance Peel in her memoir *How We Lived Then*, and related changes in the London lighting to perceptions of time and space:

> The street lamps were dimmed and no big groups or long lines of light were permitted. From Hampstead and Highgate the city could be distinguished, but there was no glare. London had gone back twenty years as regards lighting. By the end of the war it had been in the Middle Ages.[23]

The lights of London were dimmed and then darkened because of the war economy and because of the blackout regime that was enforced to protect civilians from the Zeppelin raids. Peel, however, perceived the disappearance of artificial light from the urban sphere as a journey back in time, first to the nineteenth century and then to the medieval era. This perception was not unique to her: as Lynda Nead showed in *Victorian Babylon*, London's illumination by gas, from the mid-nineteenth century, and later by electricity, was identified with the city's modernisation and industrialisation, so much so that an 1872 strike by London's gas stokers plunged the city "back into the darkness of seventy years earlier."[24] City lighting, therefore, was identified with concepts of time and progress, as reflected in the famous remark by British Foreign Secretary Edward Grey on the outbreak of the Great War: "The lamps are going out all over Europe, we shall not see them lit again in our life-time." Nead pointed to the role of lighting in the urban sphere, and noticed that "gas lighting had converted London nights into days" and that "street lamps represented the intrusion of daytime order and the rational space of the improved city into the darkness of the city at night."[25]

The threat of the Zeppelins forced a blackout regime in London, and changed the modern perceptions of 'day time' and 'night time' that had taken shape over close to a century of gas and electricity lighting. "Yes, the 'Lights of London' may be said to have been extinguished," wrote Michael MacDonagh on 31 December 1914.

> It is at night-time that we notice most the changes in our life caused by the War [...] In order to make it difficult for the enemy to find London, London, in a word, has been blacked out.[26]

The total darkness of the 'night time' was part of the urban landscape of the war. If before 1914 'day time' intruded into the 'night time', as Nead observes, then during the war the two time-frames became separated. The resulting and hard to ignore darkness served as constant evidence to the presence of war in the urban sphere. "In the evening, Sheridan and I went to the picture palace," wrote Victor Smith in his diary on 10 October 1914. "[We] were reminded of the War as soon as we came out. The blinds were drawn in the trams to dim the light in case any hostile aeroplanes should be about."[27]

Darkness became a symbol of danger – the danger of an aerial raid, and of walking in the street without seeing anything. Lillie Scales from Hornsey Lane wrote in her diary on August 1915, "Many gas lamps in the street are turned out altogether and the roads are as black as pitch [...] You cannot see the edge of the pavement, and you knock into people."[28] Like Scales, the journalist Henry William Lucy, of 42 Ashley Gardens, did not feel safe outside his house after sunset. “Whilst the authorities with increasing vigilance condemn London to a condition of Cimmerian darkness,” he wrote in his diary on 24 November 1915, “people walking in the streets at night, dodging the perils of the roadway along which the taxi-cabs and motor-omnibuses flash apparently faster than ever.”[29] Victor Smith himself was almost a victim of the darkness. “Coming home from church it was so dark, as a precaution against Zeppelins,” he wrote on 10 October 1915, “that I stepped off the pavement inadvertently and nearly went headlong in the road.”[30] And Milton Valentine Snyder, a correspondent of the New York newspaper *The Sun*, described this blackness in a letter to his wife Alice who was in Paris at the time. “I got my first impression of London by night in wartime,” he wrote on 2 February 1918 about his first night in the city:

> So impenetrable was the darkness that I did not know when we were crossing Waterloo Bridge [...] It was not till I got inside the inner lobby of the hotel that I saw light for the first time since reaching London.[31]

The new dangers of the ‘night time’ and its separation from the ‘day time’ imposed changes on the daily life in the city. “Owing to the darkness of the London streets at night and the reluctance of people to be out, the theatres are altering the hours of their performance,” reported Georgina Lee in her diary on 16 October 1914. “These now occur every afternoon instead of evening, and the ‘matinées’ on Wednesday and Saturday now take place in the evening.”[32] In addition, the Defence of the Realm Act (DORA) led to severe restrictions in pub closing times, mostly to limit drunkenness and enhance work productivity. The restrictions remained in place until the twenty-first century, and Jay Winter has described them as “perhaps the most tangible long-term legacy of the First World War.”[33]

The peril of the Zeppelins belonged to the ‘night time’ and was separated from the ‘day time’. However, not all the appearances of ‘night time’ were equally dangerous – and it was because of the ever-changing symbol of the ‘night time’ that its presence in the urban sphere became even more noticeable. “I don’t think we have any of us loved the moon as we do now,” wrote Miles in her diary on 1 January 1916. “She seems like a protecting angel shining down upon us from the sky.”[34] The airships’ pilots preferred to raid on moonless nights, when they were less visible to the British defence forces, and Londoners adapted their home-front life accordingly. “When the moon is out Zeppelins stay in, and we can go to the theatre with an easy mind,” wrote Michael MacDonagh in his diary on 1 October 1916. “Places of Entertainment have begun, in fact, to advertise ‘moony nights’ as holiday resorts do ‘sunny days’.”[35] At the front, the darkness of the night was safe cover for the soldiers, who used it to get outside the bunkers and prepare for another day of life-threatening dangers – but at the home front, night became the time of danger.

As of the end of 1916, the Zeppelins became a lesser threat in comparison to the German aeroplanes, which actually preferred a lighted landscape for their raids. The moon, therefore, turned from a symbol of safety to a symbol of danger. "O moon," wrote MacDonagh in his diary on 30 September 1917, "the enemy has enlisted you as his ally."[36] Like MacDonagh, Miles also felt betrayed. "In the Zeppelin days we looked upon the moon as our great friend and protector," she wrote on 16 September 1917. "*Now* we simply dread the moon."[37]

The betrayal of the moon – "the foul pro-German moon," as MacDonagh referred to it[38] – was another example of how perceptions of time-frames were changing because of the Zeppelins' threat. When the blackout regime stopped the intrusion of 'day time' into the 'night time', the moon was left as the sole source of light during the night. Its light, however – and later the lack of it – was recognized as a contributory factor to a successful aerial raid, and so its presence in the sky was always noticed.

When the Zeppelins came to London, the city was blacked out, but the bombs they dropped lit the streets – and interfered with the separation between day and night. "There was a hum of engines, and looking up we saw the Zeppelin," wrote 12-year-old Jacob Littenstein of 22 Orde Hall Street, in his essay describing the airship raid of 8 September 1915. "The lurid flames were licking up the sides of the bank hungrily [...] Although it was midnight it was as light as day."[39]

A reference to the bombs and fires as an intrusion of 'day time' into the well-separated 'night time' can also be found in an essay written by another schoolboy, the 11-year-old George A. Rist of 7 Cockpit Yard, describing the same raid. "I caught sight of a bomb dropping," he wrote. "We heard shrieks and cries, glass falling, and it looked day time. The flames rose to an [sic] height of thirty feet."[40] Another schoolboy, R. Beasley, used the same terminology when he described in his essay the Zeppelin raid of 13 October 1915. "I heard a tremendous bang," he wrote, "and saw a blue light which lights up the place like day-light, and the power of the light knocks you down."[41]

Light and darkness shaped the way Londoners perceived 'day time' and 'night time' during the Great War, and the threat of the Zeppelins was the main cause for this change of time perceptions. London's 'war time' and its 'night time' – identified with the perils of an aerial raid – were, as I have shown, part of the 'state of emergency' on the home front. However, as the schoolboys' essays show, when the threat of the Zeppelins became real, during and after their raids, city life was also changed at another level – when sights of war, death, and debris were forced into the world of children. The actual raids, therefore, changed the separation between time-frames in the world of children, by turning the 'night time' into a time of danger.

Part II

"The children were told to write as much as they could"

Not much has been written about the work of the educational psychologist Dr Charles William Kimmins during the Great War. Kimmins was the Chief

Inspector of Schools for the London County Council between 1904 and 1921, and for the next two decades continued to publish works drawing on his studies of children. Following his death at 91, his obituary in *The Times,* published on 17 January 1948, stated he "made a notable contribution to the understanding of the child mind during a long and fruitful career [...] It was fortunate for the education service of a vast population that Dr Kimmins recognized the fundamental necessity of focusing attention on the younger pupils in the schools."[42]

At the end of 1915 Kimmins conducted two major studies on reactions to the war by children in London. The first one was presented to the Psychological Section of the British Association on 9 September 1915, was later adapted for an article in *The Journal of Experimental Pedagogy and Training College Record* which was published on 3 December 1915, and also appeared as a chapter in Kimmins' 1926 book *The Child's Attitude to Life: A Study of Children's Stories.* For this study, Kimmins collected 3,081 essays written by 1,511 boys and 1,570 girls in ten senior departments of junior schools in London. "No preparation was allowed, and no notice given," explained Kimmins at the beginning of the article. "The children were told to write as much as they could about the war in fifteen minutes. No child was allowed to exceed the time limit."[43]

The second study was presented to the Child Study Society at the Royal Sanitary Institute on 9 December 1915. It was adapted as an article in *The Journal of Experimental Pedagogy and Training College Record*, published on 4 March 1916, and was also included in *The Child's Attitude to Life.* This study was based on 945 essays written by boys and girls, and Kimmins explained that his motivation was the satisfactory outcome of the previous study. "The results interested me so much that I thought it would be of value to carry on a similar investigation on the interests of children, at different ages, in air raids," he noted in the article. "For this purpose the children in the schools in the immediate neighbourhood of the raids of September 8th and October 13th, have written essays under the same conditions as the war essays."[44]

Both studies are cited in Susan R. Grayzel's *At Home and Under Fire* together with some of Kimmins' conclusions, and briefly mentioned in Terry Charman's *First World War on the Home Front.*[45] These are two rare references to Kimmins' interest in children's perceptions of the war and the Zeppelin raids, and both are based solely on press coverage of his two lectures. This part of the chapter, however, is based on Kimmins' articles and books, along with the press coverage of the lecture, and chiefly on the 48 complete essays written by schoolboys from Princeton Street School in Bedford Row, in London's Holborn district. The 48 essays have not been linked so far to Kimmins and were not known to be part of his study about reactions to the Zeppelin raids.

Kimmins was interested in the way children perceived the war, and in the differences between the perceptions of boys and girls at different ages. His aim, he commented, was to get first-hand information from the children themselves, and he concluded that "in the air raid stories, the nearest approach to this ideal was reached."[46] The air raids Kimmins was referring to were two Zeppelin raids on London, on the nights of 8 September and 13 October 1915. They were officially counted as the fourth and fifth raids on London, out of a total of nine raids throughout the war, but the schoolboys of Princeton Street School referred to them as the first and the second raids. For example, Alexander Fraser titled his essay about the 13 October raid "The Second Air Raid," and wrote right at the beginning:

> On Wednesday evening, five weeks after the first Air Raid, I went to bed with a light heart. I heard half past nine strike, and going to the window looked out into the black night. Then heard a loud crash, a[nd] looking up [I] saw a long cigar shaped air ship. "Zeppelin again," I gasped.[47]

The reason for the difference in the counting lies in the personal definition and perception of the urban sphere of London. The first three raids on London took place quite far from Holborn, where most of the boys of Princeton Street School lived – although the first one got as close as Whitechapel – and so they probably didn't consider them an event that occurred in London, 'their London'. Perhaps this can be seen as an example of the different ways adults and children perceive space and define urban sphere – for example, in comparison to the entry in Georgina Lee's diary after the official first raid on London. "The Zeppelins came to London last night!" she wrote on 2 June 1915. "But no one was aware of it until we opened our papers at breakfast."[48] As an adult, Lee accepted the formal definition of London, but the schoolboys were probably more inclined to define their city based on their own experiences, and found it harder to address an event they were unaware of as part of their urban experience.

Many of the essays about the Zeppelin raids are well written, and the schoolboys – some of them less than ten years old – seemed to be very eloquent and expressive. This, along with the fact that the essays were written successively in two notebooks – one for each raid – and therefore were not written by all the schoolboys simultaneously, might raise the suspicion that the parents or teachers were involved in the process, and that the boys worked on a draft and made changes in their essays before copying the final version to the notebook. There is no evidence for that, or for the opposite – and since that is the case, there is no reason not to trust Kimmins and his integrity as a psychologist and researcher. "One type of investigation [...] has not received sufficient attention, and that is the indirect method of studying behaviour by getting first-hand information from the child himself by his own record of impressions on being confronted with new situations," explained Kimmins regarding his research method.

> To secure that the impressions are really personal, and not those of parents or teachers, it is necessary to obtain such a record as early as possible after the development of the new experience. The belief of the child in

> the infallibility of the information given to him by his mother and other members of the family group, and later by the teacher, up to the age of 9 or 10 years is fatal to that faithful picture of the first impression which is of such priceless value when it finds expression.[49]

Therefore, doubting the authenticity of the essays means doubting Kimmins' professionalism, which no one at the time or afterwards seems to have done.

Another reason to believe the authenticity of the essays is their importance to Kimmins' research. Kimmins was not interested in the children's perception of space and time, but rather in their perception of the war as a whole. His conclusions at the end of his studies dealt mostly with the changes in interest from age to age, and between girls and boys – like in other studies he published at the time.[50] "In children's stories at different ages, no matter on what type of material they are based, there is approximately what may be termed a special outlook for each age, varying with the mental make-up for the particular period," he explained his interests.

> From this starting point, however, there is a wide range in the manner of treatment depending upon the ability, educational achievement, and temperament of the child. The boy's story differs in many ways from the girl's of the same age. Points strongly emphasized by the boy may be entirely omitted by the girl in dealing with stories on the same subject, and vice versa.[51]

That is why Kimmins did not refer to that difference in the counting of raids – he was more interested in the eight-year-old girl who described her dream of being married to a soldier,[52] or in the ten-year-old boy who wrote: "my father captured a German soldier, who had a nice knife. The German soldier gave his knife to my father."[53]

Despite the fact that Kimmins was not interested in perceptions of time and space, it was his method and his vision that allowed those perceptions to be expressed by the children's essays – and particularly in the 48 essays written by the Princeton Street schoolboys – that are evidence of the way the children perceived a change in time-frames because of the air raids.

"The man that was next to me had his left arm blown off"

Reading those 48 essays, it is evident that the schoolboys were very much aware of the time the raid happened: in 31 of these essays, the boy mentioned the exact hour when the attack started – although they were probably not instructed to do so, since Kimmins did not mention such an instruction in any of his articles and books – and most of the time in the first few lines. For example, nine-year-old John Sandell from 4 Johns Mews started his essay with "At half past ten on Wednesday night we had a visit from Zeppelins";[54] Albert Ling started his with "On Wednesday evening at ten minutes to eleven a first bomb was dropped in the centre of Red Lion Passage";[55] and 13-year-old James Sayers wrote in his essay that "at half past nine last night, as I was getting in bed, I heard a tumult in the streets mingled with the firing of guns."[56] This can be attributed to formalities of writing, but I would like to suggest a different reason: the attacks happened late at night, when schoolboys aged 6–14 were usually asleep or about to go to sleep, and so by forcing the children to be part of the 'war experience' the Zeppelins forced them into an adult time-frame and the adult world. This blurring of the boundaries between the 'children time' and the 'adult time' was unique, exceptional, and because of that the exact time of the raid was burned in the children's memory and accordingly mentioned in their essays.

The 'children's time' was not just the hours of daylight – it was also a symbol of safety, of peace. War was traditionally part of the adult world, and therefore part of the 'adults' time' and 'adults' space', because it took place far from home. The Zeppelins, however, brought the sights and the dangers of war into the urban sphere, forcing non-combatants to be involved in the war. Therefore, the intrusion of the 'children's time' into the 'adults' time' was mainly the outcome of the two changes in perception of time which I mentioned before: the transition from 'peace time' to 'war time', and the separation between the safe 'day time' and the dangerous 'night time'.

The blurring of the lines between the time-frames was manifested in the schoolboys' essays in several ways: acts of maturity, exposure to dangers, restrained fears and confrontation with death. Each manifestation expressed the child's understanding that he must act as an adult, because he was part of an event that belonged to the world of adults, and not to the world of children. More than that, it serves as a remarkable raw and unbiased testimony to early childhood experience of the Great War aerial raids.

This occurred, firstly, through acts of maturity that would normally have been beyond childhood experience, for example in Edward Maddox recollection of the 13 October 1915 raid: "While I was lying down on my mother's bed and got the baby to sleep, I heard a very loud banging sort of sound," wrote the ten-year-old boy in his essay. "After hearing it two or three times, I quickly called my little sister and brother up out of their sleep and I made them go downstairs and made them keep there [sic] till I came down."[57] He was not the only child who, when the raid began, took care of his little brothers – so did F. McHenry. "[I] went down-stairs to see if my mother and the children were alright," he described his reaction to the raid. "I found them in a safe place, so I stayed there."[58]

Sometimes, those children did not just help their parents, but protected them too. "I looked out of the window and saw the searchlights playing on it [the Zeppelin]," wrote James Sayers, 13, after the raid of 8 September 1915, and described his commitment to the blackout regime. "'Mother?' I said, 'put the light out.' My mother started asking questions, 'Why?' and 'What for?' Without hesitating to answer, I jumped upon a chair and in a second we were in utter darkness."[59] Albert Ling explained why he did not go out to the street to see the Zeppelin, and wrote: "I stayed indoors to look after my mother, because she was frighten."[60] And Joe Bullock, in the midst of the raid, helped an unknown mother:

> A woman who came out of a house with her children ask me would I go up in her room and get her children some cloths [sic] […] I open her door it was full of smoke and I waited until it had all gone out, and when in I got her children boots, trousers, and I put out the gas.[61]

The schoolboys of Princeton Street were also exposed to dangers, because of the Zeppelins. "When I went round to our house, I saw that a bomb had been dropped on the roof but never exploded," wrote 14-year-old Gilbert E. Bracknell of 9 Ormond Yard about his experiences during the 8 September raid. "The fuse had caught fire, but was put out."[62] Herbert Manser was also faced with the possibility of his house collapsing due to a hit by a Zeppelin bomb. "I heard a br-r-r-r and then I saw a flash, followed by a bang," he wrote after the 13 October raid. "We all ran in and stood behind the door; ready to run out if the back of the house was blown in."[63]

J. Baleh first became aware of the danger through the sounds of panic, but later saw sights of war in his neighbourhood. "I heard two or three loud bangs," he wrote after the 8 September raid. "As I looked up into the sky I thought I could see red, and about five minutes after it was quite plain. This was, no doubt a house that had been set on fire by the German airmen."[64] Like him, 13-year-old Ernest Marriage from 2 Beauchamp Street was awe-struck by the burning buildings: "We saw the Penny National Bank and The Dolphin was on fire [sic]. Flames were shooting around it, even the water looked like fire."[65]

Despite the dangers that they faced, however, the schoolboys of Princeton Street, aged 6–14, had to restrain themselves. Knowing they had to adapt themselves to the expected behaviour in the world of adults – and their gendered expectations, as boys and future men[66] – they tried, therefore, not to reveal any sign of fear. Most of them did not even mention fear in their essays, and from the few who did it can be understood that it was best to hide those feelings. "Baaang! There was another crash," wrote Jacob Littenstein, 12, of 22 Orde Hall Street after the 8 September raid, and explained that his panic was only temporary. "'Bombs and zeppelins' said my aunt. She was cool but the other women were panic-stricken […] I was shivering like a jelly but I soon got over it."[67]

Some of the boys were not ashamed of their fears. "I did not see the Zeppelin," wrote Alfred Banks about the 8 September raid, "because I was too afraid to venture out."[68] J. Sandell wrote that in the middle of the 13 October raid there was a moment when he feared for his life. "I could see the shells bursting about a couple of yards away from where I stood," he wrote. "A big crash came over our heads, and I thought for a moment that it was the end."[69] It was 12-year-old Thomas Allen of 50A Ormond Yard Street who most clearly described the gap between children's natural fears during an air raid and the calm appearance these boys tried to show in what they perceived as the world of adults:

> "Zeppelins," said my mother briefly. "What," I almost shouted, "Zeppelins over here." At the mention of these dreadful monsters I felt my knees give way. I do no attempt to disguise the fact that I was frightened. I was terrified to an enormous extent."[70]

The blurring of the lines between the 'children's time' and the 'adults' time' was also manifested through the confrontation with death – and of all the Princeton Street schoolboys, it was probably R. Beasley who saw the most terrifying sights. After meeting his father and receiving a parcel from him, he walked along Wellington Street and reached the Lyceum Theatre just as it was hit by a Zeppelin bomb. "I was knocked but I soon got up and went to rush inside the theatre, when the man that was next to me had his left arm blown off and I thought I was going to be hit," he wrote after the raid of 13 October, and described what he saw inside the theatre. "When I got to the bottom of the passage, two bodies had fell [sic] in a bit [...] When the second bomb was dropped I saw a pal in the road with his head blown off."[71]

This scenario was extreme, but not unique. Other boys were also faced with scenes of blood and death, which only emphasized the obscuring of the boundaries between the world of children and the world of adults, due to the air raids. "There was a taxi driver who was smashed to pieces outside a theatre," wrote Edward Maddox, ten, about the casualties beside the Lyceum Theatre,[72] and 12-year-old Ernest Meager from 15 Devonshire Street added: "In Exeter Street they dropped some, and wrecked a bus, and the driver's brains were blown on to a lamp post."[73] 12-year-old Thomas Allen witnessed similar damage that night, in a different street. "[I] saw a sight that fill me with horror," he wrote. "In the Strand and Aldwych a large amount of damage was done and people say that lumps of flesh were found sticking to the walls and posts."[74] And ten-year-old John Haynes from 10 Red Lion Street was in Kingsway, where he encountered a horrible sight: "On the corner was a pool of blood about one yard diameter. There were pieces of flesh almost as big as my fist. A driver of a motor-bus was killed."[75]

These expressions of mature deeds, exposure to dangers, concealed fears, and confrontations with death all show how 'children's time' spilled into 'adults' time', and obscured the boundaries between the world of children and that of adults. The Princeton Street schoolboys reacted accordingly, and saw their own experiences as belonging to the world of adults. They helped their mothers with young siblings and warned the family about the Zeppelin's bombs; they stood outside when their homes were raided, and had to watch huge fires in their neighbourhoods; they knew they must overcome their fears, and did their best to hide their feelings; they witnessed people being torn to pieces by the bombs, and confronted sights of war and death in their familiar urban sphere. However, sometimes only in one sentence out of a whole essay, they were once again children. "I saw a sight that brought an exclamation from my lips […] the National Penny Bank was wrapped in flames," wrote Thomas Allen – only 12 years old, despite his articulation skills – after the 8 September raid:

> The fire was so bright that the water playing upon it looked like fire itself […] walking along East Street [I] stumbled by chance upon a house which had been gutted by fire caused by incendiary bomb […] When I reached home I received the news that a cup of cocoa was waiting for me indoors. After drinking it I went to bed and was asleep within twenty seconds.[76]

"As We Used to Do in the Zeppelin Days"

The Zeppelins helped to change the way time was perceived in London during the Great War, and the transition from 'peace time' to 'war time' was immediate. When the aerial attacks threatened to hurt non-combatants in the urban sphere, London tried to hide itself from the raiders: public clocks and Big Ben – the most famous of them and a symbol of the city – were silenced, and so time stood still during the war time; moreover, a blackout regime was imposed, separating the safe time-frame of 'day time' from the dangerous one of 'night time'. The experience of children during the Zeppelin raids – as can be seen in the rare collection of essays from Kimmins' study – presents the outcome of this change of time-frames: young children were forced to become part of the hazardous 'night time', and saw scenes of death and destruction that had never been part of their world. The Zeppelin raids in the middle of the night changed their perceptions of time and their role in it: they had to act within the world of adults, instead of their customary world of children – and they reacted by displaying maturity and hiding their fear.

When war was declared on 4 August 1914, Hallie Eustace Miles was surprised by the new 'war time' and by the physical and imagined changes that it brought. "The whole condition of things is so strange to us all," she wrote in her diary, "for to have *London* in a state of War is not within the memory of the present generation."[77] When the war ended, she used the same word to describe the transition back to 'peace time' – in which time can once again be heard. "It is so strange and significant to hear the church clocks chime and strike again, and to hear the boom of Big Ben wafted to us after the long silence; we now notice sounds that we used to hardly *hear* before the Great War," she wrote, and moved on to explain how the 'night time' is no longer separated from 'day time', and no longer dangerous:

> When I see lights burning brightly from uncurtained windows, I feel as if we ought to ring up the police station, as we used to do in the Zeppelin days […] It is now full moon, and, instead of dreading it as our greatest danger and looking up at it with indignant eyes, we say, "What a glorious moonlight night it is!"[78]

Notes

1 The writing of this chapter, and of my research as a whole, would not have been possible without the tremendous help of my supervisor, Prof. Iris Rachamimov. I would also like to thank the International Society for First World War Studies and the organizers of the 2016 'War Time' conference in Oxford, who gave me the first opportunity to present the fruits of my research about Zeppelins and the perceptions of the urban sphere in Great War London.
2 "Holborn School-boys' Impressions of Zeppelin Raids over London, Vol I."
3 Pennell, "The Germans Have Landed," 96. The most notable example of an imagination of Zeppelins and invasion fear in Britain before the Great War is H.G. Wells's 1908 novel *The War in the Air*, which depicted a global war in which the airship played a major role.
4 Grayzel, *At Home and Under Fire*, 18.
5 Mond, "I Wish the Zeppelins Will Come Forever," 66.
6 E.g. Castle, *Fire Over England*; Morris, *German Air Raids on Great Britain*; Cross, *Zeppelins of World War I*; Hyde, *The First Blitz*.
7 Grayzel, *At Home and Under Fire*; White, *Zeppelin Nights*.
8 E.g. Gregory, *The Last Great War*; Pennell, *A Kingdom United*.
9 E.g. Becker and Audoin-Rouzeau, *14–18*; Proctor, *Civilians in a World at War*; Winter and Robert, *Capital Cities at War*, vols 1–2.
10 Pignot, "Children," 29–30, 40–44.
11 Miles, *Untold Tales of War-time London*, 13–14.
12 MacDonagh, *In London during the Great War*, 9–10.
13 MacDonagh, *In London during the Great War*, 51.
14 Rachamimov, "Camp domesticity," 293.
15 Smith, *Diary of a Young Civil Servant*, 50.
16 Miles, *Untold Tales of War-time London*, 86.
17 MacDonagh, *In London during the Great War*, 103.
18 MacDonagh, *In London during the Great War*, 102.
19 MacDonagh, *In London during the Great War*, 103; Roynon, *Home Fires Burning*, 168.
20 Miles, *Untold Tales of War-time London*, 102–103.
21 Miles, *Untold Tales of War-time London*, 87.
22 MacDonagh, *In London during the Great War*, 328–329.
23 Peel, *How We Lived Then*, 55.
24 Nead, *Victorian Babylon*, 89.
25 Nead, *Victorian Babylon*, 83.
26 MacDonagh, *In London during the Great War*, 47.
27 Smith, *Diary of a Young Civil Servant*, 50.
28 Scales, *A Home Front Diary*, 37.
29 Lucy, *Diary of a Journalist*, 223.
30 Smith, *Diary of a Young Civil Servant*, 223.
31 Snyder, *Paris Days and London Nights*, 10–11.
32 Roynon, *Home Fires Burning*, 49.
33 Winter, *The Great War and the British People*, 210.
34 Miles, *Untold Tales of War-time London*, 88–89.
35 MacDonagh, *In London during the Great War*, 143.
36 MacDonagh, *In London during the Great War*, 219.
37 Miles, *Untold Tales of War-time London*, 130.
38 MacDonagh, *In London during the Great War*, 249.
39 "Holborn School-boys' Impressions of Zeppelin Raids over London, Vol I."
40 "Holborn School-boys' Impressions of Zeppelin Raids over London, Vol I."
41 "Holborn School-boys' Impressions of Zeppelin Raids over London, Vol II."
42 "Obituary: Dr. C. W. Kimmins," *The Times*, 17 January 1948, 6.

43 Kimmins, "The Special Interests of Children in the War at Different Ages," 71.
44 Kimmins, "The Interests of London Children," 108.
45 Grayzel, *At Home and Under Fire*, 43; Charman, *First World War on the Home Front.*
46 Kimmins, *The Child's Attitude to Life*, xii.
47 "Holborn Schoolboys' Impressions of Zeppelin Raids over London, Vol II." The full names, ages, and addresses of the Princeton Street Schoolboys in this chapter are based upon the school's admission and discharge registers for the years 1911–1918, which are held at the London Metropolitan Archive. Some of the boys are cited with their name only – and in some cases, just their surname and the first letter of their first name – as it was written in their essay. Those boys are not included in the register, most probably because they were admitted to the school before 1911 and were discharged from it after 1918 – but the relevant register was not saved. "Princeton Street School – admission and discharge register for boys."
48 Roynon, *Home Fires Burning*, 115.
49 Kimmins, *The Child's Attitude to Life*, 2.
50 Kimmins, "Methods of Expression;" Kimmins, "An investigation of London Children's Ideas."
51 Kimmins, *The Child's Attitude to Life*, 17.
52 Kimmins, *The Child's Attitude to Life*, 5–6.
53 Kimmins, "The Special Interests of Children," 71.
54 "Holborn School-boys' Impressions of Zeppelin Raids over London, Vol I."
55 "Holborn School-boys' Impressions of Zeppelin Raids over London, Vol I."
56 "Holborn School-boys' Impressions of Zeppelin Raids over London, Vol II."
57 "Holborn School-boys' Impressions of Zeppelin Raids over London, Vol II."
58 "Holborn School-boys' Impressions of Zeppelin Raids over London, Vol II."
59 "Holborn School-boys' Impressions of Zeppelin Raids over London, Vol I."
60 "Holborn School-boys' Impressions of Zeppelin Raids over London, Vol I."
61 "Holborn School-boys' Impressions of Zeppelin Raids over London, Vol I."
62 "Holborn School-boys' Impressions of Zeppelin Raids over London, Vol I."
63 "Holborn School-boys' Impressions of Zeppelin Raids over London, Vol II."
64 "Holborn School-boys' Impressions of Zeppelin Raids over London, Vol I."
65 "Holborn School-boys' Impressions of Zeppelin Raids over London, Vol I."
66 Pignot, "Children," 34.
67 "Holborn School-boys' Impressions of Zeppelin Raids over London, Vol I."
68 "Holborn School-boys' Impressions of Zeppelin Raids over London, Vol I."
69 "Holborn School-boys' Impressions of Zeppelin Raids over London, Vol II."
70 "Holborn School-boys' Impressions of Zeppelin Raids over London, Vol I."
71 "Holborn School-boys' Impressions of Zeppelin Raids over London, Vol II."
72 "Holborn School-boys' Impressions of Zeppelin Raids over London, Vol II."
73 "Holborn School-boys' Impressions of Zeppelin Raids over London, Vol II."
74 "Holborn School-boys' Impressions of Zeppelin Raids over London, Vol II."
75 "Holborn School-boys' Impressions of Zeppelin Raids over London, Vol II."
76 "Holborn School-boys' Impressions of Zeppelin Raids over London, Vol. I."
77 Miles, *Untold Tales of War-time London*, 14.
78 Miles, *Untold Tales of War-time London*, 161–162.

Bibliography

Archival materials

London Metropolitan Archive
"Princeton Street School – admission and discharge register for boys," LCC/EO/DIV03/PRI/AD/001.

British Library
"Holborn School-boys' Impressions of Zeppelin Raids over London, Vol. I," 1915, Add MS39257.
"Holborn School-boys' Impressions of Zeppelin Raids over London, Vol. II," 1915, Add MS39258.
"Obituary: Dr. C.W. Kimmins." *The Times* (London), Saturday 17 January 1948.

Published materials

Becker, Annette and Stéphane Audoin-Rouzeau. *14–18: Understanding the Great War.* New York: Hill & Wang, 2003.
Castle, H.G. *Fire over England: The German Air Raids of World War I.* London: Leo Cooper, 1982.
Charman, Terry. *First World War on the Home Front.* London: André Deutsch, 2014.
Cross, Wilbur. *Zeppelins of World War I.* Lincoln: Backinprint.com, 2001.
Grayzel, Susan R. *At Home and Under Fire: Air Raids and Culture in Britain from the Great War to the Blitz.* Cambridge: Cambridge University Press, 2012.
Gregory. Adrian. *The Last Great War: British Society and the First World War.* Cambridge: Cambridge University Press, 2008.
Hyde, Andrew P. *The First Blitz: The German Bomber Campaign against Britain in the First World War.* Barnsley: Leo Cooper, 2002.
Kimmins, Charles. W. "The Special Interests of Children in the War at Different Ages." *Journal of Experimental Pedagogy and Training College Record* 3, no. 3(1915): 145–152.
Kimmins, Charles W. "The Interests of London Children at Different Ages in Air Raids." *Journal of Experimental Pedagogy* 3, no. 4(1916): 225–236.
Kimmins, Charles W. "Methods of Expression Used by London Children in Essay Writing at Different Ages." *Journal of Experimental Pedagogy and Training College Record* 3(1916): 289–295.
Kimmins, Charles W. "An Investigation of London Children's Ideas as to How They Can Help in Time of War." *Journal of Experimental Pedagogy* 4(1917): 80–87.
Kimmins, Charles W. *The Child's Attitude to Life: A Study of Children's Stories.* London: Methuen and Co., 1926.
Lucy, Henry William. *The Diary of a Journalist. Vol. 2.* London: John Murray, 1922.
MacDonagh, Michael. *In London during the Great War: The Diary of a Journalist.* London: Eyre & Spottiswoode, 1935.
Miles, Hallie Eustace. *Untold Tales of War-time London: A Personal Diary.* London: Cecil Palmer, 1930.
Mond, Assaf. "'I Wish the Zeppelins Will Come Forever': The Mixed and Contradictory Representations of the Airship Raids on Britain during the First World War" (Hebrew). *Historia* 33 (August 2014): 63–93.
Morris, Joseph. *German Air Raids on Great Britain 1914–1918.* Dallington: The Naval and Military Press, 1993.
Nead, Lynda. *Victorian Babylon: People, Streets, and Images in Nineteenth-Century London.* New Haven: Yale University Press, 2000.
Peel, C.S. *How We Lived Then, 1914–1918: A Sketch of Social and Domestic Life in England during the War.* London: John Lane, The Bodley Head, 1929.
Pennell, Catriona. "'The Germans Have Landed!' Home Defence and Invasion Fears in the South East of England, August to December 1914." In *Untold War: New*

Perspectives in First World War Studies, edited by Heather Jones, Jennifer O'Brien, and Christoph Schmidt-Supprian, 95–118. Leiden: Brill, 2008.

Pennell, Catriona. *A Kingdom United: Popular Responses to the Outbreak of the First World War in Britain and Ireland*. Oxford: Oxford University Press, 2012.

Pignot, Manon. "Children." In *The Cambridge History of the First World War: Volume III, Civil Society*, edited by Jay Winter, 29–45. Cambridge: Cambridge University Press, 2014.

Proctor, Tammy M. *Civilians in a World at War, 1914–1918*. New York: New York University Press, 2010.

Rachamimov, Iris. "Camp Domesticity: Shifting Gender Boundaries in WWI Internment Camps." In *Cultural Heritage and Prisoners of War: Creativity behind Barbed Wire*, edited by Gillian C. Carr and H. Mytum, 291–305. Abingdon: Routledge, 2012.

Roynon, Gavin. *Home Fires Burning: The Great War Diaries of Georgina Lee*. Stroud: The History Press, 2013.

Scales, Lillie. *A Home Front Diary 1914–1918*. Stroud: Amberley Publishing, 2014.

Snyder, Alice Ziska, and Milton Valentine Snyder. *Paris Days and London Nights*. New York: E.P. Duttonand Company, 1921.

Smith, Victor. *Diary of a Young Civil Servant in Westminster, 1914–1918, Volume 1*. London: H. Glaisher, 1927.

White, Jerry. *Zeppelin Nights: London in the First World War*. New York: Random House, 2014.

Winter, Jay, and Jean-Louis Robert. *Capital Cities at War: Paris, London, Berlin 1914–1919*. Cambridge: Cambridge University Press, 1999.

Winter, Jay. *The Great War and the British People*. New York: Springer, 2003.

Winter, Jay, and Jean-Louis Robert. *Capital Cities at War: Paris, London, Berlin 1914–1919, Volume 2: A Cultural History*. Cambridge: Cambridge University Press, 2007.

6 Time, space, and death

Germany's living and lost aviators of the First World War

Robert William Rennie

From 1960 to 2003, the grave of Rudolf Berthold, a decorated First World War German aviator, vanished within the fortified border between East and West Berlin. Despite the eventual dismantling of the Berlin Wall and German reunification, Berthold's resting place remained unmarked. In a century defined by unprecedented technological progress, and previously unimagined destruction in war, Berthold's body is a symbolic reflection of the complicated relationship between Germany and its First World War aviators. His turbulent life and violent death also embody a complex negotiation with time during and after the conflict.

From the advent of powered flight, aviation has represented a further compression of the human experience of space and time. This compression existed largely in the imaginations of writers, futurists, and awestruck onlookers at air shows. For most people, a first-hand experience of flight remained out of reach. With the outbreak of war in 1914, military aviation expanded this experience to the men who would serve in Germany's air service. War time in the air, however, was markedly different for military fliers. Their technologically driven, violent, and often short lives, elongated limited spans of time into longer epochs. Aviators divided along generational shifts that did not reflect their respective ages but, rather, the longevity of their service in the war and the intensity of violence surrounding certain periods.

We can best access the intricacies of time as it was lived and experienced by men like Berthold through the Greek terms *chronos* and *kairos*, which demarcate time between periods of chronological, limitless time – *chronos* – and heightened moments of decision represented by *kairos*. While those who fought might not have actively or consciously engaged with classical categories, their experiences are expressed in ways that reflect these conceptions of time and which clearly demonstrate different perceptions of time during and after the war. The conflict's unprecedented nature, both in its employment of new technology and its subsequent violence, resonated as distinct periods of time for those who served.

Chronos provided a knowable sense of structure to the war's unprecedented violence by regimenting the experience of time between and during sorties over the front lines. *Chronos*, in a First World War context, was elongated and compressed as technological advancements and an increasingly violent environment elongated progressively shorter periods of time. The violent nature of aviation, too, formed the experience of *kairos*, moments of heightened terror that required intense, mindful concentration, marked by instants of decision that meant the difference between life and death. For many fliers, the two senses of time resided in a shared space, where the tedium of long flights – a *chronos* experience – could be punctuated at any moment by intense terror – *kairos* – that demanded definitive action.

Aviation also had a powerful effect on the battlefield. For the first time, military planners could see the frontline, not from the height of a high tree or tower, but from thousands of metres overhead. The potential of powered flight proved vital almost from the outset as a continuous source of intelligence. The subsequent demand for ever faster, more lethal aircraft served only to exacerbate this sense of flight time over the Western Front. During the conflict, German aviation firms manufactured 47,931 aeroplanes.[1] The latest machines consistently transcended previous conceptions of space and time, and continued to compress the experiences of both as the war progressed. As a result, aviators who fought earlier in the war had a very different understanding of time from those who served later.

For Germany's aviators, the end of the war in 1918 shattered this understanding of time and fundamentally reconfigured their identities as fliers. The compressed sense of *chronos* collapsed into an uncertain future, and despite the dangers encapsulated within wartime moments of *kairos*, many aviators longed for knowable moments of decision and action. For men like Berthold, who died not in the war, but in one of Germany's many riots in 1920, aviation's relationship with a traumatised homeland collapsed his wartime memories and uncertain future into a new and unrecognisable present. In other words, men like Berthold remained mentally trapped in a time that no longer existed.

The broken time of post-war Germany also affected the bodies of long-dead aviators who served as markers of time, as they were re-purposed, re-buried, and rhetorically reconstructed, as either representations of an earlier time or standard bearers for a new National Socialist Reich. The most emblematic example was the body of Manfred von Richthofen – the Red Baron – who remained buried in Flanders, where he died in 1918. For many Germans, the location of Richthofen's corpse meant he was physically trapped in war time, and only his repatriation to Berlin in 1925 would reunite him with the shattered Fatherland he served during the conflict.

This chapter explores the relationship between military aviation and the experience of time on the part of Germany's fliers. The men who served in the German air service were shaped by the culture of a rapidly modernising Germany in the years before the First World War. Powered flight only heightened the sense that the world was accelerating and compressing, and the flier soon represented the potential of the individual to transcend this perilous new world. With the outbreak of war, aviation reconfigured time yet again, as the sense of compressed time only intensified as both the war and technological progression created an increasingly deadly environment for fliers. I argue that the routine of fliers – of time spent behind the lines between sorties, combined with intense moments of punctuated terror – created a different understanding of *chronos* which created generational gaps between men of the same age – defined only by their entry point into the conflict. Further, I contest that the nature of military aviation, with its focus on the destruction of the enemy, cultivated a profound sense of *kairos* – intense moments of decision that often meant the difference between life and death. Aviators, too, lived in between these experiences of time during the tedium of flying to and from targets. Finally, this chapter examines the way this regimentation of time was broken by the war's end, and how the mentalities and physical bodies of aviators remained markers of a time that no longer existed.

Military aviation and war time

Germany in 1900 was a rapidly industrialising power, and consequently experienced the changing relationship with time shared by other developed nations.[2] Stephen Kern notes that the pre-war era was defined by a new sense of the present characterised by "a thickening of its temporal length beyond a 'knife edge' between the past and future into an extended interval that included part of the past and future".[3] This reconfiguration of the experience of time created a new concept of the present, one articulated by Gertrude Stein as a lived moment that "involved streaming from the past and into the future".[4] Germany's own narrative before 1914, that of being both inherently modern and traditional, embodying both the factory floor and the *Heimat* landscape, demonstrates this complex and often contested relationship with time.[5] Powered flight represented a further technological progression from the railroad, the invention that compressed time and space in the nineteenth century.[6] For many Germans, flight embodied the next stage in the process of ever-collapsing space and ever-accelerating time, and beckoned a future where aircraft would traverse previously impossible distances in mere minutes.

The arrival of war in 1914 fundamentally changed the relationship between time and aviation. The power of invention had combined with the power of nationalism to further the development of the aeroplane at an astounding pace. With war, national lines solidified and the pre-war community of fliers: men who had sought to outperform each other in peacetime flight were called up by their representative armies. These pioneering aviators became the founding members of new military air forces. In that moment, aviation, which had been practised largely as an act of technical experimentation, transitioned from the realm of potential to the concrete reality of military necessity.

The coming of war further altered the perception of time. Kern states:

> The war contradicted such notions of an extended present on a grand scale by isolating the present moment from the flow of time. However, the

other extension of the present that we observed – a spatial extension that included a multiplicity of distant events – was dramatically embodied in the war experience.[7]

Rudolf Berthold's war diary reveals this fracture of time in his entries from the first weeks of the conflict. He notes the term 'war standard' as the dominant phrase used in military drills, and that a heightened sense of urgency, of perilous time, had arrived.[8] The compression of time, the creation of a perilous future, and the regimentation of military life often erodes any sense of reliable consistency in the present or any notion of a predictable future. A publication entitled *Physical Phenomena and the War* described war's corrosive qualities on the experience of one's pervious life at home: "Gradually, imperceptibly the images and thoughts of ordinary civilian life begin to fade; thoughts of home, wife, friends, even begin to grow dim and recede in the memory. The present, the vital present, occupies and grips the mind."[9]

Flight as *chronos*

Aviation added new delineations to the experience of war time by changing the human relationship with the battlefield. The aeroplane, that invention that compressed space and time, also revolutionised the spatial perception of the frontline – not from the height of high ground or a tall tree – but from thousands of metres above what would become a sprawling network of trenches and fortifications. Aircraft expanded the perception of the battlefield while simultaneously shrinking its scale. Miles of terrain could be traversed in minutes, rather than hours or even days. The information gathered by these aviators was invaluable. From the opening weeks of the conflict, ocular reconnaissance from aircraft proved vital to the decisions made by ground commanders. By 1915, photographic reconnaissance provided remarkably clear images of the trenches, information of vital importance to winning the war. In essence, aerial photography captured frozen snapshots of time that could be internalised as intelligence for future action.

The development of military aviation over the course of the war also reordered the perception and experience of time by those tasked with flying daily sorties over the Western Front. As early warplanes were replaced with purpose-built machines, the lived experiences of pilots, observers, and ground crews, became highly regimented and ordered. Pilots and observers flew multiple missions a day over the Western Front, mapping enemy fortifications and new trench positions. By 1915, purpose-built fighter aircraft, designed to destroy enemy reconnaissance machines, grew in number and importance.[10] With these new machines, a novel dimension of war time emerged, one solely focused on hunting and destroying enemy aircraft. In that moment, a unique breed of *Kampfflieger*, or combat flier, rose to prominence.

The maturation of the air war over the Western Front, then, created two markedly different experiences in the perception of war time. First, the military demands of acquiring daily photographic reconnaissance, ranging artillery, and later in the war the punitive bombing of infantry positions necessitated multiple sorties each day. Each mission, assigned over dangerous areas of the Western Front and, sometimes, behind enemy lines, represented a heightened sense of *kairos*. These missions represented moments of decision that could shape the course of the war and, possibly, cost the lives of the airmen assigned to carry them out. Reconnaissance flights conducted daily sorties to gather photographic data of the Western Front. Images relayed back to military intelligence noted any valuable strategic information (increased troop concentrations that could signal an impending offensive, new troop fortifications or artillery batteries, damage done to friendly trench emplacements) all factored into the decision-making process by those on the ground. The need for such information meant that sorties were divided into morning, afternoon, and evening patrols that created a highly regimented, predictable sense of time for aviators serving over the Western Front.[11] Pilots serving in both reconnaissance and bombing units, as well as those assigned to *Jasta* or 'hunting' squadrons flew with grinding regularity as the war progressed. Even by 1918, as the war turned decidedly against Germany, daily missions were still conducted routinely as the *Luftstreitkräfte* attempted to create localised air superiority in the place of regional dominance.

Second, the speed at which aviation developed during the war created an artificial sense of distinct 'eras' to the air war. These eras reached out through time, redefining the perception of *chronos* experienced by aviators. This redefinition of time was facilitated through technological improvements to aircraft. The rapid speed of technological development in aviation often meant that the next generation of military aircraft eclipsed those already in service. An example of this type of rapid progression can be found in a comparison between the Fokker *Eindecker* of 1915, which had a top speed of 87mph, and the Fokker DVII of 1918, which had a top speed of 165mph. While these details have often been noted in technical histories of aircraft, this increase in speed only intensified the experience of collapsing space and accelerating time in the minds of First World War aviators. In an instant, a once reliable machine could be rendered militarily obsolete by newer aircraft coming into production.[12] Aviation's expeditious development created a psychological shift in pilots' perception of time over the arc of the war years. In other words, whereas missions created a crisis moment, a specific point of *kairos*, in which aviators faced the possibly fatal risk of military flight, the rapid development of military aircraft redefined what 'long term' time, or *chronos*, felt like.

This elongation of time is apparent in Manfred von Richthofen's writing, when describing his encounter with aviation in 1914:

> At the time I hadn't the slightest idea what our fliers did. I considered every flier an enormous fraud. I could not tell if he were friend or foe [...] Even today the old pilots tell how painful it was to be fired at by friend and foe alike.[13]

Richthofen's definition of early aviators as 'old pilots' highlights how the violent nature of the air war, coupled with its disorientating speed of development, could, in the space of a few short years, separate, nearly contemporary Germans into 'new' and 'old' pilots. The technical rate of development in the field of military aviation over the air war's dominant years – 1915, 1916, 1917, and 1918 – created machines that not only improved drastically over time, but utterly eclipsed the technology which came before. Consequently, aviators could neatly divide the air war into the subsequent 'eras' that they either personally experienced, or that preceded their arrival on the Western Front. These delineations of time were, of course, artificial. For those away from combat, the separation of mere weeks or months into epochs appeared irrational. For aviators, however, it is apparent through their writing that each separate 'era' of the air war's development was distinct. Whereas ground combat troops would mark eras through survival, aviation added a technical delineation, that of the development and implementation of new machines, to the perception of *chronos* and *kairos.*

From *chronos* to *kairos*

The German flier Peter Supf served not with a fighter squadron, but instead with an aerial reconnaissance unit whose primary task was photographing enemy positions. Supf's experiences, as well as the source material he left behind, shows us the manner in which *chronos* and *kairos* could often blur between distinct categories in the minds of fliers. Supf's experiences are documented, not in words, but in a series of photographs arranged in an album. These images provide remarkable insight into the personal experiences of the air war for a unit tasked not with shooting down enemy aircraft, but rather photographing the front. Placed in conversation with Supf's snapshots of comrades and social gatherings are images of death and destruction: crashed aircraft, bombed out buildings, and funerals for Supf's compatriots. Their placement in such close proximity, often on the same page, denotes the influence of violence on Supf's mentality. Supf's album, which spans the outbreak of war in 1914 until the end of 1917, is marked by the destruction of enemy and friendly aeroplanes, the deaths of his friends, and the carnage of the war as viewed both from the ground and from above.[14]

The role of an aerial observer too was defined by a powerful relationship with *chronos* and *kairos.* Patrols in observation aircraft were often marked by extreme tedium, with flight crews spending long periods of time simply scanning the skies for enemy aircraft and the ground for potential intelligence. The drone of the engine, the numbing cold of the elements, and the wind howling through thick layers of flight suit had an anaesthetising effect on those who flew. It was in this space that the seemingly endless stretch of monotonous *chronos* could be punctuated in a singular moment of terror – *kairos.* A mechanical failure of the aircraft (a common occurrence during the First World War), an unanticipated attack by an enemy fighter, or an explosion of enemy anti-aircraft fire could instantly snap an aviator out of the daze of *chronos* with intense fright.

The act of capturing an image, whether in the air or on the ground, was a moment of decision in itself – an instant that captured an impermanent time in a permanent medium. Supf's photo album demonstrates a profound and complex relationship between the viewer of the album and time. Photography, more than any other medium, captures time in the past at a precise and unmovable point. It differs from the descriptive nature of text and the inherent characteristics of moving film. Time, then, is frozen in a moment of decision.[15] Supf's album captures his comrades before they are killed in combat. It also features the machines that dominated the air war in the varying epochs of the conflict. Large, underpowered machines recall the first year of the war when Supf's role as an observer would have been largely that of ocular reconnaissance. The end of the album reflects the increasing violence of the conflict, and his sense of *chronos* being reshaped, as eras within short spans of time begin to feel longer. The album then, creates a curious relationship with the past. Through the images, Supf and his comrades could, metaphorically, step back into a seemingly distant past, one that might have existed only months earlier. Aerial photographic units capture the intensely complex relationship between the aviator and time, and demonstrate the effects of *chronos* on the lived experiences of aircrews, over time or during the tedium of a single flight. *Kairos*, too, is embodied in both moments of abject terror in the air, and in the decisive click of the shutter as those who were tasked with imaging the front also captured their personal, often palpably impermanent time during a war of unprecedented carnage.

Flight as *kairos*

There is a theme, as old as the mythology of Icarus, of supreme confidence that is evident in those who became aviators.[16] The writing of German fliers exhibits a sense of aviation as a transformational rite of passage. Before becoming a pilot, Manfred von Richthofen served with a cavalry unit. Writing to his commanding officer, Richthofen asked to be transferred and, after being rebuffed at first, was eventually granted a post with the flying service in 1915. He writes in his autobiography that "my greatest wish was fulfilled".[17] In 1915 Rudolf Berthold expressed a similar emotion after he gained his pilot's licence and was subsequently relieved of the frustrations of passively observing from the air to piloting his own machine. His war diary reveals the extreme sense of confidence he held as an aviator: "As soon as it can happen, I want to be a pilot. Should my skill, my will, always be dependent on another person? Should the weakness of one person hinder my strength, which knows no barrier?"[18]

It is in Berthold's war diary that we see the degree to which aviation not only cultivated a sense of extreme personal agency in fighter pilots, but also the ways in which flight changed perceptions of time. As the war on the ground deteriorated into static trench warfare with no end in sight, the development of aircraft from fragile reconnaissance machines into weapons of war empowered the men who flew them. The skill set required to be a successful combat pilot also cultivated the notion that the aeroplane provided the aviator with a sense of individual agency that his compatriot in the mud below would never have.[19] Berthold's writing denotes an extraordinary sense of confidence that rose as he moved to take the controls of the aircraft he was assigned. That sense of agency, of power, was, of course, artificial. The power of a pilot to control his plane ultimately rested on the reliability and fragility of his machine. A mechanical failure or damage suffered in battle would render a fighter plane into little more than a plummeting coffin. From the perspective of the aviator, however, a fragile sense of agency was often bolstered by faith in one's machine, and in one's ability to master it, and from the confines of the cockpit aviation presented a chance not only to survive the war, but also thrive.

Thus, the foundational experience of the aviator, that of controlling his aircraft and fighting against another opponent, brings us to the experience of violence. Violence, too, in concert with the aeroplane further shaped pilot identity and further altered their relationship with time by creating punctuated moments of terror. Aviators like Richthofen and Berthold experienced violence in the air in an extremely heightened sense of *kairos*, where minute decisions often meant the difference between life and death. In his published autobiography, *Der Rote Kampfflieger*, or *The Red Battle Flier*, Richthofen reminds us about the difference between hunting and shooting: "My father makes a distinction between a hunter, a sportsman, and a shooter whose only fun is shooting."[20] While Richthofen's description here undoubtedly highlights his extraordinary privilege as an aristocratic flier, it also demonstrates a different conception of time. The act of shooting, from Richthofen's perspective, seems to represent a failure in capitalising on the opportunity presented through *kairos*. In other words, shooting for the fun of shooting ignores the greater moment of decision at hand, and renders the shooter as less capable of being truly decisive. A hunter, by contrast, seems to command *kairos* by slowing it down, and extracting every advantage from the moment. The analogy, presented for public consumption, illustrates the difference between an aviator who views the moment of decision with the gravity it deserves, and those who are more reckless.

Richthofen also reminds us that these experiences changed over time. After downing an opponent, Richthofen writes: "I had the feeling of absolute satisfaction. Only much later did I overcome that and also became a shooter."[21] This short statement underscores a wealth of information regarding his perception of time as the war progressed. His "feeling of absolute satisfaction", which he felt during his phase as a "hunter", was derived from his conquering of the moment of decision, of moving with precision and accuracy and downing his one opponent. He notes immediately after that he was forced, through the necessities of war, to become a shooter. This concession highlights the changing nature of war over the Western Front, as the skies became ever more populated. The other phrase, "only much later", however, shows us the extreme compression of time felt by pilots during the war. Richthofen is referencing a period of time of little more than 12 to 14 months. His use of the term 'much later' shows us how that experience of war created an elongated sense of *chronos* time, punctuated by extraordinary moments of *kairos*.

This same sense of *kairos* is apparent when revisiting Berthold's private war diary. Berthold, after the traumatic loss of his friend and observer in 1915, vowed revenge for his death. For Berthold, his deep personal loss and the subsequent need to enact vengeance transformed him into a fighter pilot, and entered him into a realm of struggle and death.[22] Berthold did so by shooting down a French Voisin biplane. Describing the attack, Berthold writes: "What happened now was a few minutes' work! [...] My machine gun began its monotonous 'tack-tack'. It was not long before the Frenchmen went over [...] emitting smoke and crashing."[23] Later Berthold describes the confirmation, by phone, of his victory. "Then came the message from the foremost troop that two downed aircraft were confirmed [...] Half an hour later, cars drove towards [the crash.]"[24]

Berthold's victory, and his description of it after the fact, highlights his mental and physical experience of *kairos*, a moment of intense violence and danger. The "few minutes' work" described by Berthold likely took a matter of seconds. Extrapolating from his narrative, it is clear that a short burst of machine-gun fire hit the engine of the French machine, sending it out of control to the ground. But the experience, that of intense focus and a heightened sense of the immediate, dangerous present, forms the foundation of his intensified memory of the event. We can sense through his entry every tense aspect of the fight: manoeuvring a machine into position against a moving target, feeling adrenaline pouring through his body while closing to a range of a few feet, opening fire, and hearing his guns over the din of his engine; all of these details were cast into high relief in Berthold's mind in an instant of extreme focus.[25] The contrasting delay, that of landing his machine and then waiting hours for ground-crew members and German infantry to confirm his victory, also expresses a sense of prolonged *chronos*, of waiting to validate the intense physical and psychological experience that just occurred in the air. There is a sense, in Berthold's language, of time not only slowing down, but of a clear sense of that deceleration.

By late 1917, aviation's development had reached a new stage of ruthless efficacy, such that it began to devour the very aviators that defined the air war's earlier years. The deaths of prominent airmen often denoted the ending of a perceived epoch within the air war's history. Max Immelmann, who rose to fame during the Fokker Scourge, died in the summer of 1916. His death was closely followed by that of Oswald Boelcke, one of the first and most famous German flying aces of the war, who died in the autumn of the same year. For those who followed, men like Manfred von Richthofen and Rudolf Berthold, the era of

Boelcke and Immelmann represented a bygone age, one comprising simpler machines and less hazardous skies. Such a narrative would continue through the rest of the air war. Werner Voss, a protégé of Manfred von Richthofen, died after a prolonged fight against the 56th Squadron of the Royal Flying Corps in September of 1917. Even Richthofen, who remained the spiritual leader of the air service and embodied the false sense of invulnerability that pilots cultivated, had been badly wounded and left to convalesce after being shot in the head by an enemy observation machine. His experience of the war too was subsequently divided between the time before and after his head wound. Rudolf Berthold was also wounded on several occasions, his worst injuries suffered on 10 October 1917 when his right arm was shattered by enemy gunfire and, barely conscious, he managed to land his machine at his home airfield. The infliction of physical wounds on airmen was a mark of *kairos*, an instant where a moment of decision spared them their lives, but inflicted a toll on their bodies. Peter Supf's photo album also ends in late 1917, as the violence in the air war increased to fever pitch. It is likely that the further compression of time, as well as the increasingly deadly environment, led Supf to simply stop documenting a war that no longer resembled the conflict he entered in 1914.

Certainly the most significant German death of the air war came on 24 April 1918, when the German Army issued the following grim report: "Rittmeister Freiherr von Richthofen has not returned from pursuit of an opponent over the Somme Battlefield. According to an English report, he has fallen."[26] The death of Richthofen fundamentally changed the mentality of Rudolf Berthold who noted the loss in his war diary:

> Now that Richthofen is dead, I am the last [of the old *Geschwader* Commanders]. It is ugly and wears one down. The death of Richthofen has been very depressing. Now I have to move on as one of the old guard. Maybe I will also be killed.[27]

Berthold's comments illustrate the degree to which German airmen identified with the era of the air war that they entered. For Berthold, who served from the early weeks of war in 1914, the summer of 1918 represented an alien time from the one he knew four years prior. As the summer of 1918 wore on, more of Germany's 'old pilots', those who had served since the 'early days' of the air war, fell one by one.

While Berthold would not be killed, he would not escape the First World War without further horrific injury. That summer, as the air war had turned decisively against the Germans, Berthold was shot down by enemy aircraft, and nearly died when his crippled Fokker DVII slammed into a house after having its control wires severed by enemy fire. Berthold, whose right arm was already paralysed from his previous injuries, was unable to use his newly issued parachute to escape. Instead, he was pulled, unconscious, from the shattered wreckage of his machine and transported to a field hospital. After dozens of injuries, this final and most serious incident grounded him for the rest of the war.

War's ending, war time continuing

For men like Rudolf Berthold, war proved the defining experience of their lives. Berthold entered military service early in life and the First World War represented a distinct space and time that, even with its horrors and loss, was a comprehensible and knowable horror in which those who were destined to survive could thrive. Berthold, then, even with his life-threatening injuries, refused to accept that his role in the First World War was over. Despite being physically absent from the conflict, the war still raged in Berthold's psyche. His sister, who served as a nurse and who would occasionally write for her brother in his war diary, noted after his injuries that he "watched with increasing concern the operations at the front and at home. Soon after his crash, he was notified that his command would be given to someone else, and that he would be assigned to another use for Germany."[28]

Berthold, a highly decorated and respected fighter pilot, would watch the war end for Germany from the confines of his convalescence. By October, Berthold, who was physically broken, was convinced that air fighting would again be a possibility for him. Once well enough to write, he corresponded with his sister, stating, "the will and the skill are still there. I want my Jagdgeschwader 2 once again."[29] By the end of October, Berthold was still recuperating from his injuries. He fumed in his war diary:

> In a few days, what powerful men have built over centuries will be destroyed. What has gone wrong? One cannot say. "Peace at any price!" So shout the people! Peace! Yes, but on what conditions? Man has lost his head. They doubted the Army [...] We are still far in enemy territory [...] O Germany, where is your national pride? It is still present in France, despite the devastation in their own country [...] You have slowly but surely undermined the good spirit of the Front.[30]

11 November 1918 marked the end of the First World War for millions, but for aviators like Rudolf Berthold the transition from war time to peace did not represent the longed-for end of war but rather the ushering in of a new and incomprehensible present.

For four years, aviation provided not only a military duty and a sense of identity, but a highly regimented, ordered, and predictable sense of time. Patrols were divided neatly into morning, afternoon, and evening sorties. Such division provided clearly defined shape to the experience of *chronos* and gave meaning to the experiences of aviators in their environment. Richthofen called the Somme river valley a most "beautiful hunting ground".[31] The air war itself divided neatly into perceived epochs and the individuals who defined them: Bloody April, the Fokker Scourge, Manfred von Richthofen, Oswald Boelcke. The Armistice of 11 November, then, did not mark a new beginning; it signalled the shattering of a knowable order: a fracturing of *kairos*, and a return to an extending and disorienting *chronos*. For Germany's aviators and, in particular, long-serving fliers like Rudolf Berthold, the previous four years had been lived almost exclusively within the realm of wartime *chronos* and an empowered sense of *kairos*. The heightened sense of purpose, of agency provided through powered flight, of certainty that their actions – no matter how limited – had affected the course of the war, was terminated with the stroke of a pen.

For Germany's aviation interests, the subsequent treaty signed at Versailles in 1919 was a seismic and devastating document, which comprehensively terminated war time and left thousands to transition into a post-war world. It single-handedly scattered the network of aviators, observers, officers, engineers, designers, and industrialists who had built German aviation over the course of the war. The Treaty of Versailles removed in a most comprehensive way the regimentation of war time provided by aviation, and disbanded the very industry that provided Germany's aviators with not only their military duties, but their core sense of identity.

The treaty enacted a series of articles designed specifically to prohibit any development in the field of aviation within Germany.[32] Article 198 stated the German armed forces not include military or naval air arms. Article 199 stipulated that, "the personnel of air forces on the rolls of the German land and sea forces shall be demobilised". For the men who served as aviators, observers, and aircraft mechanics in the First World War, military obsolescence would not translate to comparable roles in civilian aviation. Article 201 decreed that "the manufacture and importation of aircraft, parts of aircraft, engines for aircraft, and parts of engines for aircraft, shall be forbidden in all German territory". Article 201 fundamentally ended the construction of any type of aircraft or aviation related materials within the German borders. Article 202 went a step beyond manufacturing and targeted research and development by ordering the surrender of all military and naval aeronautical material to the Allied and Associated Powers. This article grouped any material even remotely related to aviation under an umbrella of expressly prohibited material. The country that had been, for the better part of the First World War, on the cutting edge of aircraft development and production was banned from continuing any work in the aviation industry. Thousands of engineers, workers, aviators, designers, and intellectuals were put out of work. For those whose lives were defined by the experience of flight in the First World War, their very identity as aviators was rendered not just obsolete, but illegal.[33]

Disillusioned fliers like Rudolf Berthold, who were highly decorated and respected combat pilots during the war, viewed a shattered present through the prism of a Wilhelmine past. It is apparent from Berthold's war diary that, as the fall of 1918 gave way to the early winter months of 1919, the war he had fought for four years was not over. Indeed, the rising spectre of Bolshevism, for Berthold, represented the next grave threat to Germany. His war diary, which began the day he joined his unit in 1914, continues until his death in 1920. This context, that of a war diary that does not terminate with the end of war, illustrates the powerful disillusionment that Berthold felt following the formal ending of hostilities. The collapse of the Army along the Western Front and the subsequent chaos at home left Berthold disillusioned with his own identity, declaring, "the uniform jacket that I have worn for so long is now defiled. How difficult it is, how hard it will be for me to go on."[34]

The search for meaning amid defeat as well as social and political chaos would elicit a range of responses from both the German public and the nation's surviving aviators. For Peter Supf, the pre-war occupation of writer would serve as a prism through which to make sense of the shattered present. The 'pilot poet' expounded upon his pre-war publications by drafting a multi-volume history on the military significance of the *Luftstreitkräfte* in which he served from 1914 to 1918. Post-Versailles Germany also created a social and political need to look back at the war, and find new meaning in those famous fliers long dead as well as those still living.[35]

For Berthold, the abdication of Kaiser Wilhelm II did not signal an end to war, but rather represented another, far more disastrous, phase of an ongoing struggle. As he reflected, "Hopefully there will soon be a reckoning for the villains who have used our misery. The Kaiser and the princes have been deposed. Incomprehensible! The human mind cannot grasp how terrible it all is."[36] Berthold, who lived his life within the span of the *Kaiserreich*, could not comprehend the chaos that enveloped Germany as war ended on the Western Front and social riots and political unrest overcame the nation at home. For a staunch monarchist like Berthold, the uprisings of 1918 and 1919 represented a catastrophic present. After initially being assigned to airfield duty within the newly formed military arm of Germany, the *Reichswehr*, Berthold soon lost his position after being ordered to shutter the airfield, as stipulated by the Versailles Treaty. Suddenly, his prowess as a pilot, even a broken one, could do little to change the course of events enveloping Germany. In April 1919, he decried the taking of Munich by a group of socialists and anarchists. Suddenly, the very heart of the Reich that Berthold had served – indeed, sacrificed his body for – was at risk. It was the singular moment that radicalised Berthold, and drove him out of the *Reichswehr* and into the radical paramilitary group, the *Freikorps*.[37]

Berthold used his status as a respected German combat pilot to encourage young men to join his own *Freikorps* unit. The right-wing anger that rankled many *Freikorps* units led Berthold and his troops into a right-wing putsch to overthrow the fragile Ebert government in Berlin. Despite being ordered to disarm and disband, Berthold and his men continued the fight. After a tense stand-off in the town of Harburg, Berthold and his men attempted to surrender, only to be attacked in the streets. Rudolf Berthold, the decorated aviator, who had survived four years of war over the Western Front, was brutally killed in the streets of Harburg.[38] His body was later found, all but unrecognisable from multiple gunshot wounds and blunt force injury to his face. His uniform had been stripped, his paralysed arm pulled from its socket. For Rudolf Berthold, an incomprehensible present – a world without a conservative *Kaiser*, a world threatened by Bolshevism and anarchy – ultimately represented a war that never ended in 1918. In that violent space, Berthold lost his life on 15 March 1920. In life, Berthold could never be re-purposed in what was, for him, an incomprehensible present. In death, Berthold transitioned from an impossible present, into a re-imagined past that would be shaped by the re-appropriation of his fellow fliers.

Aviation as a site of memory, meaning-making, and mourning

Aviation had transcended time and space in the years before and during the First World War. In the aftermath of war, however, and in a nation where flight was deemed illegal, Germans found ways of re-appropriating aviation to render a new kind of time – one marked by intense loss, anger, and fear – knowable. The spectre of death in cities like Berlin had, by the mid-1920s, become so routine as to almost fade into obscurity. Unidentified victims of violent crimes were displayed at police precincts and, according to novelist Joseph Roth, were utterly ignored by the public.[39] Yet the reburial of Manfred von Richthofen in 1925, at Berlin's *Invalidenfriedhof*, would be the site of grand public as well as political ceremony.[40]

Despite dying in April 1918, Richthofen would again serve the social, cultural, and political needs of the German people. Five years after the death of Rudolf Berthold, the Richthofen family petitioned to have Manfred's remains transferred from his grave in Flanders to their private family plot in Germany. Politics soon intervened. While the family wished to have him buried next to his brother Lothar, the German Defence Ministry, hoping to raise the profile of Richthofen's reburial, persuaded the family to locate him in the *Invalidenfriedhof* in Berlin.[41] Richthofen's reburial in 1925, however, was deliberately anchored in a highly idealised past, one that provided more meaning than fact. Germany, still searching for stability and answers following defeat in war, and the political, economic, and social turmoil that followed, latched on to Richthofen as a point of pride: a hero to the German people and an adversary still respected by his opponents.

Richthofen's death, as it was experienced in the lived present of 1918, however, was significantly different from the images propagated at the 1925 reburial. The illustrated cover of the magazine *Simplicissimus*, commemorating the death of Richthofen in May 1918, could not be starker compared with the one that marked the passing of Oswald Boelcke two years prior.[42] The *Heimat* scenes of German forests that marked Boelcke's cover are replaced with a stark, wilted silhouette of a single evergreen. Anchored into a foreboding hill is a shattered propeller and upon it sits a black eagle, overlooking a singular grave. Grey fighter planes seem to continue fighting in a rambling circle in the distance. Most importantly, there is no body, as Richthofen still lay behind enemy lines. There are no classical motifs, there is no Icarus for the masses to mourn, only the ongoing circle of violence and death, and the obliteration of the landscape.[43]

By contrast, Richthofen's second funeral in 1925 more closely resembled Boelcke's ornate ceremony in 1916. The privileged deaths of Germany's aviators were again echoed in the repatriation of Richthofen's body to Berlin. As in the case of Oswald Boelcke and Max Immelmann, the fallen aviator was afforded in death what so many of his comrades in the trenches were not: a singular, identifiable, and momentous burial. The death of an aviator represented an inherent tension between *kairos* and *chronos*, as the significance of his death marked a heightened moment of reflection, soon to be submerged in the permanence of death following his burial. In other words, the body of the aviator became a marker of time in the moments between death and final burial.

The reburial of Richthofen in Berlin offered a second opportunity to craft a new narrative about Germany's experience not only in the skies over the Western Front, but about the war as a whole. In doing so, Richthofen's reburial served to repurpose time to create a new interpretation of the past in the face of an uncertain present. This reinterpretation effectively split Richthofen in two. In his reburial and re-purposing he represented an idealised past: an era when honourable Germans jousted in the skies over Flanders; an era marked by individual heroism rather than senseless slaughter; an era when ambitious aviators seized the agency given to them by the machines they controlled. Yet he also embodied a German future that was not his to live in, that of a resurgent National Socialist Reich built on a new generation of air minded, militaristic fliers.

In what can only be assessed as a highly propagandised piece, Richthofen's youngest surviving brother, Bolko, writes, "the family agreed to this in the knowledge that the remembrance and memory of Manfred were not theirs alone, but rather, belonged to the entire German people".[44] Bolko's writing sits awkwardly between two distinct periods: the German defeat of 1918, and the Third Reich of his lived present in 1933. At once his prose both recalls the events that led to Richthofen's repatriation to Germany in 1925, while echoing the sentiments of an ascendant National Socialist Germany. Field Marshal von Hindenburg's correspondence from 1918 expressed similar emotions: "As master of the German flying force, as a model for every German man, he will live on in the memory of the German people. May this be a comfort in your grief."[45]

The contradiction of the two Manfred von Richthofens, that of the living aviator who rose to notoriety during the First World War and the new symbol of a resurgent German Luftwaffe 15 years later, elucidates the complexities of memory and, ultimately, the role of remembrance in serving not the past, but an ever-evolving present. Richthofen, the landed baron from an influential, aristocratic family, was a staunch supporter of the Kaiser. Unlike Berthold, who lived and ultimately died in the aftermath of Wilhelm II's abdication, Richthofen died before Germany signed an armistice. The timing of Richthofen's death, just before the war turned decidedly against Germany in the summer of 1918, created a malleable figure that could represent a host of cultural and political meanings.

In the aftermath of the First World War, Richthofen was largely left in the wreckage of the conflict. His body remained trapped in the events of 1918, when he fell behind enemy lines and received a military burial by his adversaries. The repatriation of his body in November of 1925 opened a conversation within German culture, one that could look back on the war and mourn both the loss of a singular figure and the collective loss of the nation. The account of Richthofen's reburial encapsulates the complexities of this conversation. His reissued autobiography recalls:

> All the bells of the small Baden city began to sound [...] And so the whole population, from the oldest man to the youngest child who could walk, came to respectfully greet Manfred's body on German soil [...] The honour guard included men who had been officers in his *Jagdgeschwader* and in the 1st Ulan Regiment. In unbroken succession Berlin's population filed past the coffin the whole day.[46]

Here Bolko's recollection of his brother's reburial is both rooted in a remembered past and in a repurposed present. The language used in his account demonstrates the tension between the two conflicting interpretations of *chronos* following the war. Onlookers no doubt welcomed Richthofen back to German soil and his railcar procession received significant attention. The description of the crowds as "unbroken", and consisting of "the oldest man to the youngest child who could walk", builds a narrative of Richthofen residing both in the past and the future. The kind of mourning discussed in Bolko's narrative shows a national reaction to the rhetorical power of a mythological hero, not the loss of a human being. Through his reburial, Richthofen fractured time, from the living world of *chronos*, to a symbolic interpretation of a future where his death, rather than his life, would serve a new generation of Germans.

The description of Richthofen's personality by Bolko in a highly idealised fashion further cements this new Richthofen in the final passages of *Der Rote Kampfflieger*. Manfred is characterised as a gifted, tenacious, and inspirational leader. Bolko writes, "Manfred put an unusual amount of energy into whatever he did from the days of his youth."[47] During the war, his character inspired his fellow fliers, "all of his subordinates in the *Jagdgeschwader* had unshakable faith in him [...] he was an example to all who followed him in war".[48] "The harder and more difficult the battles became, and the more meaningful the air battle for Germany's destiny, the greater was Manfred's own sense of responsibility."[49] In these passages, we view a Richthofen who was devoid of fear, or anger, or resentment. It is the antithesis of the writings of Berthold during the war, which often described fits of depression, sadness, and anger over the loss of fellow airmen. In a moment of rhetorical overreach, Bolko recites the Latin verse featured in Wilfred Owen's *Dulce et Decorum est*, saying "And the *dulce et decorum est pro patria mori* that his teachers once preached to him, though not always to his joy [...] became the meaning of the short life of combat that was allotted to him."[50] It is telling of the intensity of the motivations behind those who sought to re-appropriate the Red Baron that Owen's authorial irony passes by unacknowledged in Bolko's prose.

The reprinting of *Der Rote Kampfflieger* in 1933, then, reinforced the 'new' Richthofen as a symbol of a resurgent Germany. The new edition featured an introduction written by Richthofen's successor as the commanding officer of *Jasta* 11: Hermann Goering. Goering, in this new introduction, re-appropriated the Baron as an example for the German people in the new Third Reich. The language used by Goering is nationalistic and bombastic. Far from the cold, resigned prose of the newspaper accounts reporting Richthofen's death in 1918, Goering states, "21 April 1933 marks the fifteenth anniversary of the day in which Rittmeister Freiherr von Richthofen, at the zenith of his glory, met a hero's death."[51]

Thus, Richthofen was reanimated rather than remembered. The German air hero who returned in the 1920s and 1930s bore little resemblance to the cold, calculating pilot who died as the First World War turned against Germany in 1918. In his place was a proto-Nazi, bent on securing Germany's place as a leading world power. Goering states:

> We will hold Manfred von Richthofen as a great symbol. His memory will help us to use all means in our power to reach our national goal of again giving Germany an air weapon equal to those of other nations, but superior to them in spirit and courageous sacrifice, as was the *Jadgeschwader Richthofen* in the World War.[52]

Richthofen himself would likely have not recognised the warped figure he became as Germany grew ever more militaristic in the 1930s. But certain aspects of Richthofen's character made his re-appropriation by National Socialism easier to achieve. Richthofen's love of military discipline, his desire to hunt his victims, his view of war as sport rather than violence, and his ability to inspire confidence in his subordinates all resonated with the radical right. By removing his human characteristics, his fear of death, and the injuries he suffered in combat, National Socialism easily recast the monarchist as a torchbearer for National Socialism. Thus, the memory of the Richthofen who died in 1918 was re-purposed to serve the needs of Germany in 1925, and again in 1933.

During the same period, Rudolf Berthold, that other flying hero buried at the *Invalidenfriedhof*, was similarly reanimated, in a changed form, by the Third Reich. Streets in Bamberg, his hometown, bore his name as he was extolled as yet another proto-National Socialist by a regime bent on re-writing the history of its fallen Kaiserreich aviators. Berthold then joined the ranks of other deceased First World War aviators. Like his re-purposed fallen fliers, Berthold was not remembered for his vehement defence of the Kaiser, but rather for his achievements in the air: that of destroying British and French aircraft in the name of the Fatherland.

The reach of National Socialism also extended to Peter Supf, who moved further to the political right. In 1935, he published the *Book of German Aviation History* in two volumes. Here again, the spectre of Goering appeared in the foreword, mirroring his words in the opening to Richthofen's reprinted autobiography. Supf then combined his experiences as an aviator and observer over the Western Front with his innate passion for writing. During the Second World War, he went on to publish histories extolling the success of the new German Luftwaffe as it devastated the European continent immediately after the outbreak of the Second World War. Thus, Supf's two identities mutually reinforced each other, and served a popular demand – at least among those of right-leaning, National Socialist ideologies – for works that celebrated Germany's prowess in the field of military flight.

Conclusion

New technologies often transcend our perception of space and time. The ascendancy of powered flight at the beginning of the twentieth century furthered the compression of space and time that began with the railway a century earlier. This transformation of time is best accessed historically, through the prism of *chronos* and *kairos* as a method of demarcating chronological time and heightened moments of decision that were often experienced by aviators. Military flight added a highly regimented sense of time, where daily sorties required constant and precise work, military aviation created a profoundly exhilarating but brittle identity for Germany's fliers, one which struggled for meaning outside of powered flight. *Chronos*, then, was redefined within the confines of an incredibly dangerous environment, where months represented epochs and the tediousness of a long flight could be punctuated by extraordinary moments of *kairos*, where minute decisions could mean the difference between life and death.

The further development of military aviation, in the form of the fighter plane, gave pilots an extraordinary sense of agency in a war marked by growing individual powerlessness amid incomprehensible carnage. An analysis of both public and private writings reveals the degree to which aviators sensed and reacted to different experiences of time. Manfred von Richthofen's hunting discourse, in addition to illustrating great privilege, also reveals a command over time, and the ability to heighten one's senses to such a degree that perceived time slows down. From Richthofen's perspective, doing so provides an advantage within the moment of decision. Rudolf Berthold's private words also illustrate the ability of some fighter pilots to slow down time by heightening their sense of it. The "few minutes' work" of his machine guns contrasts starkly with the colourless time spent waiting for the confirmation of his success within his moment of *kairos*. Violence further shaped the perception of time among aviators. This violence was only intensified by the technological evolution of aviation during the war. The rapid rate of aviation's development was so disorientating as to radically elongate the chronological experience of war for Germany's aviators. In a war that occupied just four years, pilots could be defined as 'new' and 'old' with mere weeks or months separating the two. Thus, lifespans could prove correspondingly abbreviated.

The end of the First World War shattered a highly regimented, organised, and knowable, if horrific, existence for the fliers of the German *Luftstreitkräfte.* The Treaty of Versailles, signed the following year, systematically dismantled Germany's aviation industry, and with it a relatable understanding of time for the nation's fliers. The end of the war, coupled with the end of flight, left men like Berthold disorientated and angered. By throwing himself into the social and political chaos of 1919, Berthold clearly thought that the war of 1914 never ended. In that sense, Berthold abandoned his identity as an aviator and transitioned to the only comparable role that remained, that of soldier. In repurposing his identity, he served the *Freikorps* against the next wave of 'threats' against the Fatherland. His incomprehensible present, that of a world devoid of a Kaiser and under perpetual, existential danger, ultimately cost him his life. Peter Supf, the 'poet pilot', used his writing skills to draft a history of the German *Luftstreitkrafte.* And those long dead, like Manfred von Richthofen, were resurrected both physically and rhetorically. As Berthold passed from the world of the living, he too became what Richthofen would soon represent to a new generation of Germans: a good National Socialist.

Germany's First World War aviators, then, occupied a different space of experience and memory from their fellow Germans. Aviation harnessed both technological promise and peril, creating machines that could achieve individualistic success while potentially killing those who flew. War hardened the identities created by powered flight. It was a mode of existence that ceased to exist with the armistice of 1918, and it was a world rendered extinct with the signing of the Versailles Treaty a year later.[53] When processed through the prism of aviation, the meaning of the war splintered into a multiplicity of interpretations by the living: of a struggle that never truly ended, of an epoch that demanded authorial intention. For the dead, re-purposing was perhaps easier, as Germany reanimated rather than remembered its fallen aviators to make sense of the loss. Men like Richthofen and Berthold, then, embodied the great paradox of Germany's relationship with its First World War aviators: these fliers simultaneously represented an idealised past and an unfulfilled future. It is perhaps fitting that Berthold's body remained trapped in an unmarked grave – situated in a "neutral" zone between East and West Berlin – for four decades of a divided Germany. Berthold's physical purgatory is emblematic of Germany's historical relationship with its First World War aviators: a generation of fliers who lived and died within a nation that re-imagined them, reburied them, and re-purposed them, but never fully realised their lived, horrific, and often tragic reality.

Notes

1 Morrow, *German Air Power in World War I*, 202.

2 Matthew Jefferies writes extensively on German imperial culture and his work informs my argument here.

3 Kern, *The Culture of Space and Time*, 294.
4 Kern, *The Culture of Space and Time*, 294.
5 Alon Confino's work on the nation as local metaphor (*The Nation as Metaphor*, 1997) informs my argument here.
6 My work draws heavily Wolfgang Schivelbusch's work on railroads.
7 Kern, *The Culture of Space and Time*, 294.
8 Berthold, Persönliches Kriegstagebuch, 2.
9 Kern, *The Culture of Space and Time*, 294.
10 John Morrow writes extensively on the development of the German Air Service in the First World War.
11 Weather remained an unpredictable factor in daily patrols, with weather fronts sometimes scrubbing all scheduled sorties for days at a time.
12 This placed combat aviators in a precarious position. A new, superior machine provided an overdeveloped sense of agency in the air, whereas an outdated aircraft left aviators feeling a complete loss of that agency.
13 Richthofen, *Der rote Kampfflieger*, 25.
14 Supf, Photo Album, LR-02118.
15 Dan Magilow writes extensively on the use of photography as a historical source of analysis.
16 Fritzsche, *A Nation of Fliers*, 100. I agree with Fritzsche's interpretation that "airmen emerged as representatives of a more tough-minded, popular patriotism that was technologically able and ruthlessly chauvinistic", but would further stress that aviation provided an extraordinary sense of agency to those who flew.
17 Richthofen, *Der rote Kampfflieger*, 24.
18 Berthold, Persönliches Kriegstagebuch, 13.
19 Richthofen, *Der rote Kampfflieger*, 57.
20 Richthofen, *Der rote Kampfflieger*, 103.
21 Richthofen, *Der rote Kampfflieger*, 103.
22 Berthold, Persönliches Kriegstagebuch, 40.
23 Berthold, Persönliches Kriegstagebuch, 42–44.
24 Berthold, Persönliches Kriegstagebuch, 44.
25 The machine guns on First World War aircraft were often located just a few inches in front of a pilot's face. Thus, firing the weapon would have been incredibly loud and violent.
26 Richthofen, *The Red Baron*, 141.
27 Kilduff, *Iron Man,* location 2105 of 4085 [eReader edition].
28 Berthold, Persönliches Kriegstagebuch, 63.
29 Kilduff, *Iron Man*, section 2391 of 4085 [eReader edition].
30 Berthold, Persönliches Kriegstagebuch, 65.
31 Richthofen, *Der rote Kampfflieger*, 55.
32 Norman Graebner and Edward Bennett write extensively on the Treaty of Versailles.
33 John Morrow offers a wider assessment of Versailles on Germany's aircraft industry.
34 Berthold, Persönliches Kriegstagebuch, 66.
35 Jay Winter rightly notes that the scale of grief following the First World War was such that the belligerent nations' surviving populations felt compelled to look back, rather than forward, to assuage their grief.
36 Berthold, Persönliches Kriegstagebuch, 66–67.
37 Berthold, Persönliches Kriegstagebuch, 82.
38 Berthold, Persönliches Kriegstagebuch, 106.
39 Black, *Death in Berlin*, 19.
40 Black, *Death in Berlin*, 20.

41 Lothar von Richthofen was also a famous ace in World War I, and served alongside Manfred in *Jasta* 11. He died in a flying accident in 1922.
42 Fritzsche, *A Nation of Fliers*, 99.
43 Fritzsche, *A Nation of Fliers*, 99.
44 Richthofen, *The Red Baron*, 151.
45 Richthofen, *The Red Baron*, 142.
46 Richthofen, *The Red Baron*, 155.
47 Richthofen, *The Red Baron*, 163.
48 Richthofen, *The Red Baron*, 165.
49 Richthofen, *The Red Baron*, 166.
50 Richthofen, *The Red Baron*, 166.
51 Richthofen, *The Red Baron*, 183.
52 Richthofen, *The Red Baron*, 183.
53 Germany, of course, found ways to circumvent the restrictions of Versailles.

Bibliography

Unpublished materials

Berthold, Rudolf. Persönliches Kriegstagebuch. MSG2–10722. Freiburg Bundesarchiv. Supf, Peter. Photo Album, LR- 02118. Archiv der Deutsches Museum, Munich.

Published materials

Black, Monica. *Death in Berlin: From Weimar to Divided Germany.* Cambridge: Cambridge University Press, 2010.

Confino, Alon. *The Nation as Local Metaphor: Württemberg, Imperial Germany, and National Memory, 1871–1918*. Chapel Hill: The University of North Carolina Press, 1997.

Fritzsche, Peter. *A Nation of Fliers: German Aviation and the Popular Imagination.* Cambridge, MA: Harvard University Press, 1992.

Kern, Stephen. *The Culture of Space and Time: 1880–1918*. Cambridge, MA: Harvard University Press, 1983.

Kilduff, Peter. *Iron Man: Rudolf Berthold: Germany's Indomitable World War I Fighter Ace.* London: Grub Street, 2012.

Morrow, John. *German Air Power in World War I.* Lincoln: University of Nebraska Press, 1982.

Richthofen, Manfred von. *Der rote Kampfflieger.* Berlin: Verlag Ullstein & Company, 1917.

Richthofen, Manfred von. *The Red Baron*. Translated by Peter Kilduff. New York: Doubleday & Company, 1969.

7 The photo albums of the First World War

Composing and practising the images of the time of destruction

Erica Grossi

Over the course of the First World War, the use of photography at the front intensified and its products were mass-diffused. It played a major logistical role in the conduction of military actions, as well in assuring media coverage of the European trenches for the benefit of the home front.

However, photography bypassed these official functions due to its capacity to freeze time in its frames, to suspend the "dangerous moment" in the *click* of the photographic shot. Indeed, the German word *Augenblick* indicates both 'instant' and 'blink of the eye';[1] in the same way, the French word *instantané* may mean 'something instantaneous' and 'snapshot'.

The number of pictures produced on the Western fronts by the official military photographers, as well as by the *amateur* soldiers, led the National Military Photographic Sections to create strategies to collect and conserve them. At this point in time, one of the most functional and widespread objects in use was the photo album, already well known since the middle of the nineteenth century as a way to preserve the visual memory of the family.

Through this type of illustrated book or visual memoir the war can be displayed, showing the materiality of events through images as well as activating and reactivating the memory of that modern destructive time through the 'snapshots' that materialise it. These photo albums seem to activate, in every collective, national, and individual 'present', the material, fragmented, and suspended past of the First (World) War Time of the twentieth century: an event that the French historian Marc Bloch, in 1921, defines as "one immense experiment in social psychology".[2]

Drawing from several Italian examples, this chapter will analyse the anthropological role of the album as an optical device for familiarising the human gaze with "the destructive character" of Modernity,[3] and for composing and practising the visual materiality of historical time.[4]

First World War and photography: transdisciplinary research on the physiognomy of modern time

> It is likely, that as long as I live, at least if I do not become senile in my last days, I shall never forget the 10th of September, 1914. Even so, my recollections

> of that day are not altogether precise. Above all they are poorly articulated, a discontinuous series of images, vivid in themselves but badly arranged like a reel of movie film that showed here and there large gaps and the unintended reversal of certain scenes.[5]

This is how Bloch describes the functioning, through "a discontinuous series of images", of the memory of his experience at the front during the First World War. Starting with the words of this historian is not only a way to underpin the theoretical premises of this chapter. On the contrary, from this and other passages of Bloch's reflections on the First World War, one can deduce his suggestion to look for a methodology that – "for the sake of our science",[6] the historical research on this conflict – may be able to interpret all the aspects of this modern 'event-experiment'.

During and immediately after the war, there was a widespread interest in objects and documents different from those traditionally analysed in the international academic field of First World War studies. The newly founded *Kulturwissenschaften* (cultural studies) – in the language of its founder, the German art historian Aby Warburg – considered every human production, in every time, as a cultural expression, and thus as a source for the interpretation of events, cultures, and societies on all levels. Amongst those who experienced and simultaneously commented on the war, several scholars and artists were drawn towards the discourse and the methodological strategies connected to the visual sphere, especially cinema and photography.[7] In their war productions, especially the photographic creations, they exploited these artefacts in order to reconstruct and historically represent the experience of millions of people involved in the conflict.

One of the most interesting production is Warburg's *Kriegskartothek*, a private collection of images and press materials created between 1914 and 1919. This archive is the main source for the publication of the two editions of the Italian *La Guerra del 1914: Rivista illustrata* (*The War of 1914: Illustrated Review*),[8] and for the composition of some panels of the *Bilderatlas Mnemosyne.*

Warburg perceived immediately the tragic and immense catastrophe of the war, and he focused his attention on media, in particular visual productions produced and diffused by the national propaganda services. His 'war collection' also includes several pictures from the European fronts of the conflict. In this way, stuck on the 'home front', Warburg tried to collect and conserve the visual memory of the events, composing a personal archive with the images taken by "different eyes". At the same time, he also used those picture on the pages of the *Rivista illustrata,* to show 'visually' the horrible mistake of that war. In Warburg's opinion, aesthetics and visual culture are the pivotal fields of research to understand the human experience and to express its perception of reality. The intensification of the use of photography in the trenches and the mass diffusion of its products suggested to Warburg the peculiar affinity between visual devices – cameras, *montage*, album, illustrated book, and press, atlas, etc. – and the destructive physiognomy of the modern First World War time.

This example – even if briefly introduced – underlines two pivotal issues in the reflection on the Italian photo albums of the First World War. On the one hand, it shows the familiarity of the nineteenth and twentieth centuries' mass culture with photo-cinematographic techniques and productions; on the other, it encourages scholars to apply a new perspective when engaging with the photographic and iconographic archives of this conflict.

We shall add to these aspects another element that is fundamental to grasp the complexity of the objects of this chapter: among Bloch's scientific and documentary heritage we can find also a series of pictures (approximately 109), which the French historian possibly took, and definitely gathered, during his experience in the trenches. It would not be odd, in light of what he says in his memoirs, to consider them as an attempt to bring order to the blanks and discontinuities of his memory. These images are collected in two albums, the first consisting of 67 pictures, the second of 42. In the first album it was the historian's mother who created the album, with the pictures that Bloch gave her and commented upon during his leave from the front. The captions she added following Bloch's comments are not dated, "as if the war was a long and unique episode".[9]

In the second album, instead, the captions are written by Bloch himself: they often bear dates and report some information, more frequently short and incomplete but sometimes rich in detail. If, as some commentators wrote, the first album allows the reader to enter the space of war, the second adds time and, in this specific case, that of the historian.

Actually, if we confront Bloch's statements with his photographic production from the front – but it is not known if Bloch owned a camera or if he just collected pictures from his comrades – the addition of temporal details in the captions is not sufficient by itself to restore a definitive chronological order to what happened in the trenches. What Bloch's albums truly stress is that the photographic device works analogously to the perceptual modalities of the aesthetic experience in those trenches. In both contexts the events manifest themselves through bursts, shocks, and explosions, through alternating flashes of light and darkness, through discontinuous accelerations and zooms, through intermittent visions in a field of view delimited by the frame of the trench. Snapshots fixed on the memory the physiognomy of that catastrophic time as images on the pages of a photographic album.

These opening considerations enable us to underline the three major aspects on which this chapter will focus:

1 the methodological potential of pictures as documents of the First World War – their mass diffusion allows us to consider them as visual evidences of the "everyday life [...] fragmentation that the war time makes apparent at the level of public conscience";[10]
2 the narrative strategies and mnemonic functions activated by the anthropological experience of the photographic album – this shall be considered as a technique to "impose order on the first great disorder of the world";[11]

3 the extended catalogue of 'figures' – characters, objects, landscapes, and ruins – that in the albums of the First World War act as icons that uncover the familiarity of the European West with "the destructive character" of the twentieth century of the World Wars.

Since the 1970s, the new methodological and transdisciplinary approaches of cultural history have developed a growing scientific interest towards the participation of the masses in the First World War. This has led many European scholars to adopt a 'marginal' and 'from below' perspective compared with the traditional military and politico-diplomatic historical approach. Literature on this topic is boundless and it would be impossible to make a thorough list in this context, but as an example it shall be sufficient to consider two of its most recent publications: *Encyclopédie de la Grande Guerre 1914–1918. Histoire et culture* [12] – and its translated editions – and the European digital project *Europeana 1914–1918*. Along with an attention towards the egodocuments and small artisanal objects produced in the trenches, the interest of these scholars focuses also on the peculiar relation between war and photo-cinematographic devices, especially in France by philosophers, historians such as Annette Becker (2014), Grégoire Chamayou (2015), Georges Didi-Huberman (2009), Laurent Véray (2008), and Paul Virilio (1989), and international scholars such as Peter Burke (2002), Nicholas Mirzoeff (2005), William J.T. Mitchell (2011), and Susan Sontag (2003). In Italy this approach has developed along two tracks: on the one hand, cultural history – by Antonio Gibelli (2002), Mario Isnenghi (1980), Gian Piero Brunetta, Giovanni De Luna (2006), and Angelo Schwarz (1980); on the other, visual and media studies – by Andrea Pinotti and Antonio Somaini (2016), and Maurizio Guerri (2012).

Amongst the archival collections, the great importance of pictures and photographic albums is representative of the 'mass-nature' of the experience in the trenches. Paradoxically, despite their documentary value, these sources have until recently remained under-explored.

If we say, as the French school of the *Nouvelle Histoire* does, that each age produces sources that respond to its needs and desires,[13] then photographic albums show an evident capacity to deal – quantitatively and qualitatively – with a specific need. Indeed, the First World War soldier, limited to the trenches, experienced a new space-time condition in which aesthetic perception became integral to the war strategy. The visual framing of this *Homo Spectator* [14] *sub specie bellica* is both augmented by new optical instruments and reduced to the tiny and hidden embrasures of the underground observatory decks. In the logistic of the 'trench warfare', the perception is thus completely different from that on past war scenarios, even those of the end of the nineteenth century.[15] This new perception is not only strategically important since it allows a better observation of the surroundings and a further degree of visibility on the field, it is also fundamental to 'aiming' and 'taking a shot' at targets. The overlapping of the gun-sight, the viewfinder, and their positioning in the framing of the trenches are fundamental in establishing the character of the testimonials collected through the pictures of the First World War.[16] Photographic albums are capable of gathering up these scattered pieces of the war puzzle as it is fragmented in millions of 'dangerous moments': the snapshots taken on all the fronts of the war.

The extensive use of the photo camera – both by the official photo-cinematographic services of the armies and the individual soldier equipped with amateur portable cameras – is a response to an essentially temporal need. In the trenches, indeed, the soldiers experience a time both dilated and slowed down and that may be abruptly accelerated or shaken by blasts, sudden actions, and unexpected explosions.[17] These 'dangerous moments' in which the constant menace of death coexisted with the tedious wait for the catastrophe is materialised and crystallised in the photographic click; that is the shutter click that immortalises the blink caused by a sudden bang.

The analysis produced in the fields of history of philosophy, cinema studies, visual culture, and media theory proposes an interpretation of this massive diffusion of photography as an aesthetic practice aimed at the precariousness and the pitiable instability of life during the First World War. One of the cultural goals of photography between 1915 and 1918, especially amateur photography, is to secure in a material object the fragile time of the soldier's existence: the presence of his physical body in the shaky field of the history of the first global conflict.

The compilation of these pictures into albums and its fruition satisfies a series of specific historical needs, aimed at giving an order to the chaos of the world. First, these activities give a material possibility to reconstruct the landscape, fragmented and divided by the trenches and the military walkways. Secondly, they allow the disposition of objects and people in spatial and temporal contiguity in order to make them recognizable once again. Thirdly, they give them a new readability in a war time in which everything seems to have lost its meaning. Lastly, the album propagates photographic codes as pictorial *aide-mémoire* to the telling of these stories. "When the (scattered) pictures take sides"[18] on the pages of the album, the agencies implied are twofold: on the one hand, the individual subjects that fabricate their own private and intimate visual stratification of the memory of war – as in the case of Bloch and his mother, for instance; on the other, the military institutions that dispose the pictures collected on the fronts to visually recompose the different phases of the war to exploit them in the context of a public use of the past.

Before presenting some tangible examples of these objects marked by "the destructive character" of the First World War time, it seems appropriate to further elaborate on the history of the photographic album. More specifically, we will explore the moment when the album meets the "terrible love of war"[19] of photography, becoming its most trusted companion in the cultural sphere.

The album as an illustrated book of time *sub specie bellica*

Since its invention, the use of the object-album has to do with a need strictly connected to the one that founds the epistemology of the historical discipline and of all its *ancillae*. The use of Latin proves instructive when exposing the history of an object whose primary goal, in modern times, has been to collect and preserve memories.

The word *album*, whose etymology is commonly connected to the inventions of photographic albumin, actually originates from the Latin adjective *albus* (white), which in ancient Rome indicates a little white board used to write down official acts, private and public announcements, and legislation. Therefore its function, since the beginning, is both an informative and normative one, a kin of *pro memoriae* in the guise of registration lists.[20] This lasted until the sixteenth century, when the album progressively turned into a book of personal memories becoming, especially in the eighteenth century, a bound volume meant to collect autographs of friends and loved ones (*album amicorum*).

Gradually the expression ended up designating almost exclusively variously structured collections composed by objects of different natures organised in a unitary context. The album becomes a repository of memories: letters, antiques, drawings, *cartes de visite*, all that is needed to compose the memory of its owner. It is a private museum in which the objects live in space-time continuity and contiguity, even when the bonds that keep them together are not immediately evident.

At this point of development, the reasons that made the album a successful and widespread device become clearer. To the same extent, it is possible to understand the further popularity gained by this 'memory container' once its history meets that of photography. Following Michel Foucault, the eighteenth and nineteenth centuries are the *tournant* at which Western civilisation develops an obsession for history:

> the will to enclose in one place all times, all epochs, all forms, all tastes, the idea of constituting a place of all times that is itself outside of time and inaccessible to its ravages, the project of organising in this way a sort of perpetual and indefinite accumulation of time in an immobile place, this whole idea belongs to our modernity.[21]

This need takes the form of the accumulation of traces and objects that document the passing of time as well as a historical consciousness of it. Hence the development of theoretical basis and scientific methodologies for the study of 'histories', as well as the edification of facilities and normative basis to organise the traces of time in archives, museums, and libraries, privileges locations in which the forms of time are filed *ad infinitum*.

In this phase the album had the form of a portfolio: its boxy shape progressively evolved into that of an actual book on whose pages the first photographic plates are glued. The actual "Pictorial Turn"[22] in the use of the album takes place in 1839, with the invention of the calotype, named after its creator William H. Fox Talbot. It allows to reproduce the same image in several copies and to perfect it serially on sheets of albumined paper – the egg's albumin is used to fix the image. This technique explains the familiarity of the word *album* with the practice of pictures' reproduction through the use of the albumin and the subsequent collection of the albumined papers in photographic books.[23] From this moment on, instead of collecting objects and samples on site, it is sufficient to take pictures and then paste them on the pages of the book-album to gather information and memories of exotic or scientific journeys around the world.

As it happens for photography, the employment of the album by the natural and applied sciences led to its significant diffusion and innovation. The second half of the nineteenth century witnesses a parallel proliferation: on the one hand, the diffusion of photographic plates employed for scientific experimentation and archaeological and naturalistic research; on the other, the practice of album as a *medium* of classification, comparison, and communication of knowledge. As a didactic and scientific device, the photographic album allows the conservation and the repeat of documents in locations and moments very distant from the ones in which they are acquired. The album's organisation follows a principle of selection and composition aimed at the fruition of images and the gathering of knowledge on the portrayed objects. In this sense, it represents at the same time a device of photographic world-making that activates the scientific notions and a more accessible tool for the circulation of this knowledge.

In addition to this informative and didactic character, since its origins the album has shown an evident propensity for what we call today 'intermediality'; that is the interaction in the same device of elements that differ in terms of typology, support, and interpretative codes. In our case, all these components concur to the effectiveness of *montage* and to the definition of the album as a collection. It appears that the earliest examples of the albums resemble more than others those that portray the experience of the First World War due to the presence of captions, maps, graphs, numbered lists, tables, newspaper clippings, or military dispatches; for instance, the *Album photographique de la mission remplie en Egypte par le Vte Emmanuel de Rougé de l'Académie des Inscriptions et des Belles-Lettres accompagné de M. Le Vte de Banville e de M. Jacques de Rougé attachés à la mission 1863–1864*, published in Paris in 1865;[24] or the Eugène Sévaistre albums on the siege of Gaeta by the Garibaldi's Mille and on the Palermo Revolution of 1860.[25]

During the second half of the nineteenth century, with the increased possibility of mass reproduction of photography, the album lost its exclusive nature of collection of souvenirs to become a new potential instrument of *mise en livre.* Its new congenial bookshape led the album-collection to acquire a more complex space-time structure, gathering simultaneously topographic and temporal traces of the experiences. This alliance between book-shape and photographic *medium* is strengthened by the truth-value that photography acquires at the end of the nineteenth century. This time is indeed characterised by the positivistic ideology of progress applied to everyday life, including the optical and photographic devices and strategy. It concurs to radically improve their documental value and, more than anything, the perception of their historical nature.[26]

The modern logic of 'accumulating the past' finds in the compilation of an album a strategy to bring order in time through the disposition of the photograms. Indeed, in this device, the 'temporal theme' is a fundamental point of analysis. It has a double character and multiple functions. First, it is a tool to preserve, make use and activate the past while reordering it. This is done through the anthropological act of compiling pictures that are in themselves devices to 'immortalise' fractions of time. This verb adopts from the Latin root of the word *mors, mortis* (death) the meaning of 'not being subject to death' and 'to perpetuate the memory'. The more common use of this verb refers to the photographic practice: to immortalise, thus, stands for to photograph.

Secondly, the album offers the possibility of time travelling in a growing number of directions – being lobbed back-and-forth by its cross-references, the quality of images it contains, and the point from which the consultation starts. In this sense, as a system organised by the invisible connections between images, maps, tickets, and captions, the album is a Bakhtinian 'chronotope': "a fictional fusion of space and time".[27]

This complex network of suspended relations technically defines the device of *montage* as the sequencing of photograms of time fixed in snapshots. In this sense, the adoption of the concept of chronotope is justified since it reveals the complexity of different temporalities activated by its peculiar spatial structure. In other words, the album is a chronotope insofar as it is a device for the conservation of time in a spatial order and, simultaneously, it produces, activates, and re-activates the multiple temporalities preserved in the framing of the single pictures.

The importance of *montage* for the photographic album has three fundamental reasons. The first is that, despite the book-shape assumed by the album, its structure follows an ordering logic based on seriality, simultaneity, and adjacency, which is thus not necessarily linear. The second reason deals with the material possibility of fruition: the visual act of leafing through the pages, choosing or following an order and a direction. It deals especially with the impulse of 'leaping over' given by the spaces between the images, those empty intervals that recall Bloch's description of the discontinuity in the images of memory. Ultimately, the photographic *montage* upsets the 'traditional' readability of historical time, which, in the album, is visually narrated by the simultaneous presence of different images in the same page and by their arrangement in a grid which is rarely regular or symmetrical.

It is, above all, the contiguity of these images as a 'chronotopographic' characteristic of this narration that, along with the other features, makes the album a device particularly apt to represent the chaotic experience of war. Furthermore, it responds to the discrete attempts, private and collective, to rearrange the snapshots in an order capable of filling the blanks of memory and of bringing 'tellability' to the photographic hoard.

This narrative, mnemonic, and organisational nature of the photographic album explains its huge success and diffusion in the material and visual culture of the European metropolis between 1890 and 1910: the 'family album', as the most efficient tool to preserve, for private use, a visual memory of an intimate story. A gender/class/race-specific activity, the album is composed mainly by women and finds its place in the drawers of the bourgeois parlours, as Benjamin points out in his theory of the late-nineteenth interior.[28]

Since its massive diffusion as an illustrated book of 'photo-graphic' words – the "words of light" as Talbot called them[29] – this "exhaustive inventory of the visible world", capable of reducing its historical complexity to an "accessible format"[30] gains a new feature. The album becomes a "continuously evolving" container of visible similarities, a kind of kinship that excludes what is invisible and extraneous. Each family album proposes and fixes through 'time-frames' the 'visible figures' of the invisible parental bonds. This is why, both anthropologically and aesthetically, anyone who leafs through any of these albums can sense the "family likeness".[31]

At this stage, this 'cultural familiarity' of the spectator with the objects, publications, and contexts in which the photographic albums are produced becomes a familiarity with the device itself and the *montage* logic of compiling, ordaining, and conserving. This familiarity – interpreted both as the proper cultural practice of the family group and as an expression of a shared atmosphere – is the first aspect that shall be considered to understand the next phenomenon analysed in this chapter: the photographic albums produced in the atmosphere of the First World War.

The Italian photo albums of the 'Fondo Guerra': immortalising the experience and monumentalising the war

In his *Little History of Photography*, Walter Benjamin writes that the nature speaking to the camera is different from the one speaking to the human eye.[32] Following this intuition, we might understand the reason why from the press services to the avant-garde artistic productions created during and just after the war, from the official archives to the private ones, the photographed nature speaking to the photographic eye kept on re-proposing a huge scenario of war.

During the First World War, photography portrayed reality as a vast stretch of ruins, corpses, militarised landscapes, wire fences, gas masks, and *gueules cassées*. If, on the one hand, photography becomes a logistic, documentary, and propagandistic tool to relate the different war events, on the other, the visual culture that originates from it absorbs entirely and seamlessly this same militarised catalogue of the world.

Indeed, while the war becomes 'total' because of its geopolitical extension, duration, and engagement of human and material resources, the potentialities of the mechanical reproduction of photography define visually "the destructive character" of the new century that arises with that conflict. As the De

Goncourt brothers predicted: "Everything is becoming black in this century: photography is like the black clothing of things"[33] and of the humanity "in the dark times."[34] In the progression 'through shocks' that is the history of the twentieth century as a battlefield,[35] the psychological and social landscapes are populated by objects, images, and aesthetical gestures *sub specie bellica*, as to demonstrate *ad oculos* that the world is fully ready for its destruction.[36]

Between the end of the nineteenth century and the beginning of the First World War, the technical image became the principal tool to document scientific experiments and research. This is shown by its first massive employment as a tool to document war during the first European colonial conflicts in North Africa and in the Middle East. In this context, photography sustained (pseudo)scientifically the colonial efforts by portraying the enemy as inhuman and barbarian in opposition to the civilised and enlightened people from the West. The *longue durée* of this experimentation on the battlefield produces what, drawing from Benjamin, some cultural historians of the twentieth century called "cultural brutalisation".[37]

As for the European visual culture between the two centuries – and particularly the Italian one, which is the focus of this chapter – the socio-political emergence of the bourgeoisie, the economic and cultural enthusiasm for scientific progress, and rapid industrialisation put photography to work to construct a strictly aesthetic knowledge on the world. The camera of the reporter captured train crashes and scenes of urban crime but it reached also the more charming and exotic 'Oriental' sceneries, precisely where the European troops engaged in the last peripheral conflicts before 1914–1918.

This globalisation of photographic culture, which precedes the Great War, progressively familiarised the civilian spectators with the brutal visions of the first massacres carried out with modern weaponry. Indeed, if we briefly go through the history of the relations between photography and war, we might recognise the signals of a mutual and progressive subjugation. It starts with the Crimea War (1854–1856) and establishes itself progressively through the American Civil War (1863–1865), the Paris Commune (1870–1871), and the Russo-Japanese War (1904–1905). Adopting a Foucauldian terminology, we may define this phenomenon as 'war panopticism': the conflict is exhibited everywhere and photography is employed both for the construction of a war imaginary and for the management of the war itself.[38]

In the *longue durée* of this mutual enchantment between photography and war, the years of the First World War represent a turning point and show a further increase in the diffusion of this technique. Its systematisation responds to the militarisation of everyday life, rather than to the global nature of this conflict. The will behind the diffusion of such a large quantity of images may have been intended as both official and private: the purpose was to make the war known on the home front through the efforts of the photographers enlisted in the Photo-Cinematographic Services of the armies. In addition to this, the first pocket cameras also allowed the *amateur* soldiers to experiment with this practice from their front position, increasing with their pictures the magnitude of the archives of the official services. Photography thus thrives in the trenches, as much to document the military operations and everyday life at the front as to compose a vast war album meant for mass cultural consumption.

Between 1914 and 1918, all the nations engaged in the war created teams devoted to the photo-cinematographic documentation of military activities. The first country to act in this sense was Germany with the creation of the 'Militärische Film und Photostelle' in 1914. Several months later, in the spring of 1915, the French government established the 'SPA – Section Photographique de l'Armée', later transformed in 'SCA – Section Cinématographique de l'Armée', under the direct supervision of the Ministry of War.[39] For the Italian context, the creation of the 'Servizio Fotografico del Regio Esercito Italiano' (Photographic Service of the Italian Royal Army), which at its peak employed about 600 professional photographs, led to the production of approximately 150,000 plates and films, and, consequently, to the constitution of a documentary heritage consisting of millions of pictures from all the war fronts and from every different point of view, subjective perspective, and anonymous gazes.[40]

This extraordinary amount of photographic collections needs to be considered in its entirety and also in its anonymity, as a collective and multifocal documentation that needs to be ordered and compiled to be publicly usable. Between 1915 and 1919, the task of collecting the pictures produced on the Italian war fronts was carried out by the 'Ufficio storiografico per la mobilitazione industriale' (Historical office for industrial mobilisation)' – since August 1916 under the jurisdiction of the Ministry of Weapons and Ammunition that echoed the Jüngerian *Mobilmachung*, the 'total mobilisation' of the European nations between the beginning of the century and the end of the war – and by 'Comitato Nazionale per la Storia del Risorgimento Italiano' (National Committee for the History of Italian Risorgimento), created before the war under the jurisdiction of the Ministry of Public Education. Immediately, these official bodies committed to the collection – right on the fronts or in the immediate rears – of "single images, later pasted in dedicated albums with some handwritten indication that would serve as captions".[41] All this was then duplicated and reproduced by the 'Archivio dello Stato Maggiore dell'Esercito' (Army High Command Archive).

This collection system went on until the 1930s, following the nationalistic discourse of fascism that depicted the First World War as the fourth Italian independence war against the Austrian occupant. It is not a case if the 'Fondo Guerra' (War Archives) – that included also the albums, the loose pictures, the drawings, the propaganda posters and the illustrated press – was meant to be conserved in the national monument that celebrated the 'father of the nation', King Vittorio Emanuele II, and the 'martyr son', the Unknown Soldier. The 'Istituto Centrale per la Storia del Risorgimento' (Central Institute for the History of Risorgimento) – consisting, today, in the archives, a library, and a museum – was thus assimilated into the structure of the 'Vittoriano' of Venezia Square in Rome. Between 1934 and 1937, along with the suppression of the 'Comitato' and the application of the Giovanni Gentile reform law of the Italian institute of culture, the collection was dismembered and divided.

The 'MCRR – Museo Centrale del Risorgimento' of Rome (Central Museum of Risorgimento) keeps 122 photographic albums, identified on the basis of the original indications on the front page or on the first inventory label. Conserved as 'Fondo Fotografico della Prima guerra mondiale' (First World War Photographic Archives), it is composed of 50 albums made by the 'Sezione Fotocinematografica del Regio Esercito Italiano' (Photo-cinematographic Section of the Italian Royal Army); three albums of honours for the Unknown Soldier; two albums dedicated to China and Japan (1905–1907); 15 albums of special reprints hailing from the armies of the central empires; ten albums on mountain warfare; five albums on Aeronautic; four albums belonging to Vittorio Emanuele III. The remaining are either albums made by a single photographer whose name is indicated on each compilation or albums dedicated to post-war reconstruction. As Langford writes about the albums in the McCord Museum: "Ironically, the very act of preservation – the entrusting of an album to a public museum – suspends its sustaining conversation, stripping the album of its social function and meaning."[42]

As far as the temporality of photographic production is concerned, the accumulation peak was between 1917 and 1918, following the *débâcle* of Caporetto, even if the first 'Comitato Nazionale' call for documents for the constitution of a war archive dates from August 1915, when the Minister Paolo Boselli issued the official communication.[43]

At present, on the occasion of the centenary, the proliferation of European and international projects aimed at the digitisation and web sharing of the First World War's archives led to the creation of the official website www.14-18.it (for its page specifically for the albums: www.14-18.it/album), on which the Italian collections, along with many others, are stored.

The photographic albums are thus the outcome of a specifically oriented policy of picture selection and of specific temporal scans articulated in their *montage*. The endless possibilities offered by this composition practice, along with the logistic and documentary purpose, open new possibilities for the interpretation of photography as a source on different levels of the experience of time and of "the destructive character" of war. Furthermore, the centenary constitutes a commemorative and celebrative occasion that, on a European and global level, represents an interesting opportunity for historians to question the familiarity with which these peculiar sources are usually approached.

The material time of destruction: on some Italian iconic photograms

In the context of a reflection on 'First World War and time', the photographic album represents a major methodological potential of it – it can create a sense of familiarity between the eye of the today's observer and the collective military imaginary represented photographically from the beginning of the twentieth century until the *Interbellum*. Indeed, we might say that the First World War initiated a new aesthetic character, that "destructive character" that was going to be the brand of the twentieth century. For this reason, in this chapter we consider the albums as tools to visualise the 'photo-genetic' forms of this militarised age that begins in 1914.

The pictures collected in these albums depict reality only as a succession of terrible battles and events of destruction, leaving aside everything that does not conform to this militarised ambiance. Millions of snapshots of this kind flooded European visual culture for over four years, mass diffusing this representation of wartime. This phenomenon revealed – while producing it – an aesthetical congeniality between photography and the catastrophic face of the war that would become, as already said, the original character of the twentieth century. This congeniality between photography and war scenes produces a psychological and aesthetical mass phenomenon of familiarity in the spectator who starts to 'recognise' this kind of visual imaginary. This relation seems defined by a sort of a Wittgensteinian 'family resemblance' that permeates in a similar way all the photographic albums of this war, exposing the same vast and familiar modern world *sub specie bellica*.

As previously said, the photographic *montage* has to be understood as a cultural practice that answers the anthropological need for conserving time-fragments in time, thus as an essentially historical activity. In the specific case of the experience in the trenches, the *montage* suggests an epistemological coherence between the structure of the album itself and the puzzled reality seen by the gaze framed and amputated by the embrasures on the battlefield.[44] Indeed, the interwar period is marked by the election of the illustrated book and of the *montage* as experimental forms to restore the 'readability' of the world after the estranging events of the First World War and the tragedy of modernity.

As previously said, several figures of European cultural world experimented – both artistically and epistemologically – with *montage* as a technique of production, alteration, and interpretation of the temporal complexity they lived because of the war. The experience of the trauma of the technological, phantasmagorical, and pyrotechnic war suggested and developed an aesthetic *penchant* for the temporal simultaneity expressed by the typographic-optic-synoptic configuration as well as in the estranging time of film and the shocking time of advertisement.

As for the photographic albums, the *montage* allows us to read the single time-fragments conserved in the images in a synoptic context – the album's page – in which simultaneity and seriality alternate, intersect, and re-compose. As for the album's spectator, this practice generates several different impressions: the landscape becomes a 'total' war territory; every human figure embodies a different version of the same mobilised soldier; the nature portrayed by the photographic eye is a repository of similar ruins, twisted iron fences, burned shrubs, and dark ruts on the ground, indistinct piles of snow and greyish moon-like peaks.

These pictures of the time between 1914 and 1918, arranged in albums serially produced and now also virtually accessible as a form of 'Historytainment' of the First World War, are an experimental excavation pit in which we can find some "deep figures"[45] of the imaginary of the war that endure in the visual culture of the twentieth century. These forms not only populate the single picture but, as 'figures' of the wartime they characterise it aesthetically. In other words, they crystallise the first great destructive manifestation of the twentieth century in its perceivable material forms.

The Italian case appears to be particularly interesting due to the fact that it articulates these 'figures' in a series of different and peculiar ways. If, on the one hand, this specificity reveals the variety of possible worlds *sub specie bellica*, on the other, it inscribes itself firmly in the two major iconic time-related categories of the visual materiality of the wartime that will be analysed as a conclusion of this chapter: the "faces of the soldiers"[46] on the warfront and the "inscription of war"[47] on the face of the world.

In the first category, we can consider, as Jünger suggests in his pseudo-scientific catalogue of human profiles of war, the possible 'types' of configuration of humanity as 'cannon fodder', a human material among the others in the 'war of materials'. This aspect can be analysed – quantitatively and qualitatively – in the Italian case of the 'Fondo Fotografico' because of the specificity of the Italian mobilisation for the war. For the first time in national history, leafing through the pages of these albums we discover the faces of the Italian men, as if it is an immense national 'family portrait'.

As already said, the First World War sent to the battlefront millions of men from the countryside as well as from those social classes that experienced the nineteenth century from the doors of the factories that were changing the rhythms and forms of everyday life. It was first of all the uniform that transformed these different men into interchangeable elements placed on the battlefront.

In the specific chronotope of the Italian albums, this standardisation process is incremented by the successive re-elaboration and picture cataloguing operated during fascism, when the ideology of the nation is radicalised. This element, already strongly present in the Italian and European propaganda since the beginning of the war, was indeed reinforced through the fascist discourse on the unanimity of the blood and body of the patriotic armies fighting for the final unification and liberation from the Austro-Hungarian menace. The 'monumentalisation' of the photographic albums in the same marble altar dedicated to the first king of Italy, along with the body of the Unknown Soldier, sets in a definitive way the collective 'family resemblance' of the Italian nation in the millions of faces and profiles conserved on the pages of these volumes. These icons – massively and serially reproduced on leaflets, small holy pictures, in the illustrated press, and on books published also in peacetime – constitute, even morphologically, the definitive image of the patriotic body of the family-nation. This image, born and modelled photographically in the trenches, was then monumentalised and so it became eternal.[48]

In the second category we classify all those elements that, in the photographic composition of the album, may be identified as inscriptions of the wartime on the face of the earth: permanent physical and cultural signals of the action of the conflict in the surrounding reality. As for the Italian case, its peculiarity lies in the predominance of images of mountain landscapes, the Italian experience being one of a "vertical war".[49] "The destructive character" that elsewhere, for example in Flanders fields or on the French plains, cut horizontally and in depth through the surface of the world, in Italy took shelter on mountaintops[50] and hid in caves.[51]

In the Italian landscape, the ruins fixed on the pages of the albums are represented by the same peaks, deformed by watchdogs and buttresses.[52] The specific natural monumentality of this landscape testifies in a visually eternal way to the devastating action of the human war. In other words, the ruins that elsewhere have been reconstructed, preserved, or demolished are literally inscribed in the 'faces' of these mountains as a kind of visual "natural history of destruction".[53] From the visual point of view, they are also a white and uniform presence, perfectly in a *pendant* with photography's two-colour print and with the chromatic coherence of the interstices between the frames of photograms and the backgrounds of the albums' pages.[54] In the Italian context, the background becomes a photogenic subject in itself, a natural landscape completely militarised. In these mountain pictures we recognise elements of domestication *sub specie bellica* (trenches, wire fences, weapons, and war instruments) of nature conquered in the patriotic effort of the war. These specific objects, as previously said, take the place that, in other pictures, is occupied by the ruins of buildings and architectural structures of destroyed cities and villages.[55]

Italy discovered its mountains as they became, during the war, the fixed background of the nation's fourth war of unification. The landscape, endlessly repeating itself in the albums, marked by the white tones of the perennial snow – the Italian war being also a 'white war' – and the silver shades of the photographic surfaces[56] became 'photo-genetic' subject of the long cultural history of the "homeland mountains".[57]

At this point of the analysis, it seems interesting to propose a peculiar Italian intersection between these iconic categories: the case of the 'Alpini', the special mountain troop of the Italian Army. If we look at the albums dedicated to their challenges, we can figure out how the vertical and white landscape became the most iconic scenario for the military actions of this corps.[58] In the pictures, the 'alpino' soldier and the mountain at war live in a symbiotic way:[59] the first embodying the Italian hero and the national sportsman of a kind of athletic record-war; the second one – the mountain – monumentalising the epic roughness of the Italian war efforts.[60]

In this sense, these two icons 'figures' with their multiple forms prove to be the most 'symptomatic' instruments in order to detect the conditioning of the visual culture concerning the First World War in Italy.

> The geographic position of the Alps along the boundaries of Italy evoked a defensive conception of war that went hand in hand with the

> celebration of the fallen. The Alps were the geographical sentinels of the nation, the keepers of the memory of Italians rising in arms to defend their homeland. War and politics thus redrew the map of Italy. [...] Nature had transfigured warfare and the soldiers and then preserved the traces of war; politics organised and put an official stamp on that memory.[61]

This peculiarity of the Italian album's archive accustoms the spectators' gaze to human and landscape profiles often traced in the snow and living in the temporal ambiguity of the 'mediatic aura' of photography. Enveloped in whitish opacity of perennial snow, they are presented as actual crystallisations of time.[62] The same effect of space-time 'iconisation' is produced, in the pages of these albums, by those pictures that propose futuristic, dystopian, and sci-fi views of the landscape,[63] postcards of astral territories and lunar rocks.[64] As in the case of group pictures and 'genres scenes' in which the subjects strike poses in order to appear and correspond to the ordained 'family resemblance' of the album, even the photograms of war abide by a specific iconographic code that is still lasting nowadays in the new conditions of serial fruition.[65]

Conclusions

In conclusion, we can identify three main ways in which these photographic albums relate to the issue of 'First World War and time'. On a first level, the photo album, as a device to collect fragments of time – the *instantanés* – is a material container of historic events as a discontinuous series of moments-images. Consequently, the war photo album is also a peculiar illustrated book enabling today's spectators to have various and always varied epiphanies of what the First World War is in terms of visual experience. As mentioned above, visual perception is the main turning point registered in the human experience during the 'state of emergency' of the First World War, due to the use of advanced logistic technology, including photography, to visualise the world *sub specie bellica*.

Finally, on the Italian perspective, we can consider these photographic albums as various catalogues of what – during and immediately after the war – was going to become the most powerful 'figures' to construct, both aesthetically and emotionally, the rhetorical tale of the victorious 'armed Homeland'. As already said, the albums show the portrait of a country represented by the sacrifice and athletic struggle of these hikers-soldiers, immortalised and eternalised in the Italian visual memory. It is this same 'time phenomenon' that also affected the image of the nation as a new 'land of conquest': on the one hand, the Alps are a wild border territory that represent both an athletic conquest and a geo-political victory on the Austro-Hungarians; on the other, for the first time, the photo albums allow a view of the Italian people as a militarised community struggling to unify once and for all not only its so-called *irredento* (unredeemed) territory, but also its society as a 'photographed family' feeding its "imagined community".[66]

If we broaden our perspective beyond the Italian case to the photographic album collections of other countries, we find ourselves facing the same situations as the one studied for the scientific, archaeological, and family albums. As much as the single photograms from the opposed battlefronts of France, the central Empires, or the colonies display dissimilar situations and report different information and sceneries, in the complex of this kind of heritage the 'family resemblance' is similar, syncretic, and, indeed, familiar. In other words, the nature speaking to the camera reveals visually "the destructive character" of wartime. The catastrophe has an intrinsic quality: it is equally photogenic on each ruin and human profile. This 'talent' transforms these images in icons of the Western family album of war, as recognisable as Auschwitz's barbed wire or the atomic mushrooms.[67]

In conclusion, the goal of this chapter is to strongly encourage scholars and communities to shed new light on these albums so that, as relics and heirlooms, they shall pass from the silence of the archives to public sharing, as everyone does, from time to time, in their home with their family albums.

Notes

1 Jünger and Bucholz, *Der gefährliche Augenblick.*
2 Bloch, Réflexions d'un historien sur le fausses nouvelles de la guerre," 18.
3 Benjamin, "Der destruktive Charakter."
4 Archivio del Museo Centrale del Risorgimento di Roma (MCRR) Album P 7(1). All MCRR images reference in this chapter are accessible in digitised format online by following the hyperlink for each image listed in the bibliography.
5 Bloch, *Memoirs of War, 1914–15*, 89.
6 Bloch, "Réflexions d'un historien sur le fausses nouvelles de la guerre," 35.
7 I.e. the aforementioned Warburg and Bloch, Walter Benjamin, Ernst Jünger, Ernst Friedrich (1924), the Dada artists John Heartfield and Kurt Tucholsky (1929), László Moholy-Nagy (1925), etc.
8 All translations are mine unless noted otherwise.
9 Becker, "Préface", XIV–XV.
10 Catucci, *Per una filosofia povera,* 224.
11 Didi-Huberman, "Échantillonner le chaos".
12 Audoin-Rouzeau and Becker, *Encyclopédie de la Grande Guerre, 1914–1918.*
13 Le Goff, *La nouvelle histoire.*
14 Mondzain, *Homo Spectator.*
15 Virilio, *War and Cinema.*
16 Grossi, "La memoria nelle fotografie dalle trincee della Grande Guerra", 196–205.
17 Gibelli, "Nefaste meraviglie", 547–551.
18 Didi-Huberman, *Quand les images prennent position.*
19 Hillman, *A Terrible Love of War.*
20 Quetin, "L'album photographique", 5–52.
21 Foucault, "Of the Other Spaces: Utopias and Heterotopias", 26.
22 Mitchell, *Picture Theory.*
23 Le Corre, "Les albums de photographie: une lecture dirigée", 19–25.
24 In Gallica digital collections: http://gallica.bnf.fr/ark:/12148/btv1b84583622.r=album%20photographique?rk=901292;0.
25 Partially available on the website of the Civico Archivio Fotografico of the city of Milan.

26 Crary, *Techniques of the Observer*; Sicard, *La fabrique du regard.*
27 Langford, *Suspended Conversation*, 44.
28 Benjamin, “Paris, Capitale du XIXe siècle”, 60–77.
29 Becker, *Voir la Grande Guerre*; Cadava, *Words of Light.*
30 Rouillé, “Éditorial: Vertige du nombre”, 5.
31 Didi-Huberman, *L'Album de l'art à l'époque du Musée Imaginaire de Malraux*, 26.
32 Benjamin, “Little History of Photography”, 510.
33 Becker and Philips, *Paris and the Arts, 1851–1896*, 40.
34 Arendt, *Men in the Dark Times.*
35 Traverso, *L'histoire comme champ de bataille.*
36 Patočka, *Heretical Essays in the Philosophy of History.*
37 Banti, *L'onore della nazion*, 352.
38 Grossi, “La violenza nelle sperimentazioni visuali”, 207–219.
39 “La section photographique de l'Armée et la Grande Guerre”, 210–217
40 Fabi, *La prima guerra mondiale 1915–1918.*
41 Gioia, Pizzo, and Santiemma, “Ricordando la prima guerra mondiale,” 69.
42 Langford, *Suspended Conversation*, 5.
43 Pizzo, “Fondo Fotografico della Prima Guerra Mondiale”, 288–355.
44 MCRR Album G Grande(91a).
45 Banti, *Sublime madre nostra*, VI–VIII.
46 Jünger, *Das Antlitz des Weltkrieges.*
47 Farocki, *Images of the World and the Inscription of War.*
48 MCRR Album S 17(40.21).
49 Leoni, *La guerra verticale.*
50 MCRR Album B 1(29).
51 MCRR Album B 1(81).
52 MCRR Album B 1(39); MCRR Album B 1(33).
53 Sebald, *On the Natural History of Destruction.* See image MCRR – Album Fp 3(150).
54 MCRR Album M 1(1).
55 MCRR Album O 1(6).
56 MCRR Album M 1.
57 Armiero, *A Ragged Nation.*
58 Mondini, *Alpini. Parole e immagini di un mito guerriero.*
59 MCRR Album M 1(9).
60 MCRR Album M 1(13).
61 Armiero, *A Ragged Nation*, 92. See image MCRR CA 4(2597).
62 For example, *La Guerra Bianca*, National Geographic Italia, 2016, photographs by Stefano Torrione, www.stefanotorrione.com/portfolio_page/la-guerra-bianca/.
63 MCRR S 17(7).
64 MCRR CA 4(2594).
65 For example, see the records for the ‘alpini’ research on online First World War archives: www.14-18.it/ricerca?searchFld=alpini&searchType=simple&paginate_pageNum=1.
66 Anderson, *Imagined Communities.*
67 Traverso, *L'histoire comme champ de bataille.*

Bibliography

Unpublished material

Archivio del Museo Centrale del Risorgimento di Roma (MCRR) – Fondo Guerra – Album fotografici http://www.14-18.it/album

Album P 7:
MCRR P 7(1) www.14-18.it/img/album/mcrr_p_7/full.
Album G Grande:
MCRR G Grande (91a) www.14-18.it/img/album/foto/6590/full.
Album S 17:
MCRR S 17(7) www.14-18.it/img/album/15360/fa4819fa1b8bb5161b1f65810b843425/full.
MCRR S 17(40.21) http://www.14-18.it/img/album/mcrr_s_17_40_21/fotografia/0021/full.
Album B 1:
MCRR B 1(29) www.14-18.it/img/album/15029/9e85e2d85309fdfe979ca9133fe733ba/full.
MCRR B 1(33) www.14-18.it/img/album/15029/721ed7d8edf65be1117e97ee877011 0e/full.
MCRR B 1(39) www.14-18.it/img/album/15029/6d02ebe458d818b15230773eccc5e6b2/full.
MCRR B 1(81) www.14-18.it/img/album/15029/7fc4a9af213e9efb7d9c3b236d90a213/full.
Album Fp 3:
MCRR Fp 3(150) www.14-18.it/img/album/mcrr_fp_3_p150/full.
Album M 1 www.14-18.it/album/16934/2b9a79ae67040c94a83e7396a98da16f :
MCRR M1(1) www.14-18.it/img/album/16934/2b9a79ae67040c94a83e7396a98da16f/full.
MCRR M 1(9) www.14-18.it/img/album/16934/29006b1502647ef410e0a27c9b0a4f34/full.
MCRR M 1(13) www.14-18.it/img/album/16934/10637a52ecb6bfc6c676c845477b4c66/full.
Album O 1:
MCRR O 1(6) www.14-18.it/img/album/mcrr_o_1_p006/full.
Album CA 4:
MCRR CA 4 (2594) www.14-18.it/img/cartolina/MCRR_CA_4_2594/1/full.
MCRR CA 4(2597) www.14-18.it/img/cartolina/MCRR_CA_4_2597/1/full.
Civico Archivio Fotografico – Civiche Raccolte Grafiche e Fotografiche, Lamberto Vitali/LV12; LV 25. http://fotografieincomune.comune.milano.it/FotografieInComune/fondi/FON-3a010-0000004.
Civico Archivio Fotografico of the city of Milan. http://fotografieincomune.comune.milano.it/FotografieInComune/dettagliofotografia/LV_00012_001?context=photoByAuthor&position=1&contextRefValue1=Sevaistre,%20Eug%C3%A8ne; http://fotografieincomune.comune.milano.it/FotografieInComune/dettagliofotografia/LV_00025_279?context=photoBySearch&position=316.
Europeana 1914–1918. www.europeana.eu/portal/it/collections/world-war-I.
Bibliothèque Nationale de France (BNF) – Gallica. http://gallica.bnf.fr/accueil/?mode=desktop.

Published material

Anderson, Benedict. *Imagined Communities: Reflections on the Origin and Spread of Nationalism*. London: Verso, 1983.

Arendt, Hannah. *Men in the Dark Times*. New York: Harcourt, 1968.

Armiero, Marco. *A Ragged Nation: Mountains and the Making of the Modern Italy: Nineteenth and Twentieth Centuries*. Cambridge: The White Horse Press, 2011.

Audoin-Rouzeau, Stéphane, and Jean-Jacques Becker. *Encyclopédie de la Grande Guerre, 1914–1918: Histoire et culture*. Paris: Bayard, 2004.

Banti, Alberto M. *L'onore della nazione: Identità sessuali e violenza nel nazionalismo europeo dal XVIII secolo alla Grande Guerra*. Torino: Einaudi, 2005.

Banti, Alberto M. *Sublime madre nostra: La nazione italiana dal Risorgimento al fascismo*. Roma-Bari: Laterza, 2011.

Becker, Annette. "Préface." In Marc Bloch, *L'Histoire, la Guerre, la Résistance*, edited by Annette Becker and Étienne Bloch, VII–LX. Paris: Éditions Gallimard, 2006.

Becker, Annette. *Voir la Grande Guerre: un autre récit (1914–2014)*. Paris: Armand Colin, 2014.

Becker, George J., and Edith Philips. *Paris and the Arts, 1851–1896: From the Goncourt Journal*. Ithaca: Cornell University Press, 1971.

Benjamin, Walter. "Der destruktive Charakter." In *Illuminationen: Ausgewählte Schriften I (1920–1940)*, 289–290. Frankfurt am Main: Suhrkamp, 1977.

Benjamin, Walter. "Paris, Capitale du XIXe siècle: Exposé de 1939 – écrit directement en français par Walter Benjamin." In *Das Passagen-Werk, Band V, Teil I*, 60–77, Frankfurt am Main: Suhrkamp, 1982.

Benjamin, Walter. "Little History of Photography." Translated by Edmund Jephott and Kingsley Shorter. In *Selected Writings: Vol. 2, 1931–1934*, 507–530, Cambridge, MA: The Belknap Press of Harvard University Press, 1999.

Bloch, Marc. "Réflexions d'un historien sur le fausses nouvelles de la guerre." *Revue de synthèse historique* 33, no. 97–99(1921): 13–35.

Bloch, Marc. *Memoirs of War, 1914–15*. Translated by Carole Fink. Ithaca: Cornell University Press, 1980.

Burke, Peter. *Eyewitnessing: The Uses of Images as Historical Evidence*. Ithaca: Cornell University Press, 2002.

Cadava, Eduardo. *Words of Light: Thesis on the Photography of History*. Princeton: Princeton University Press, 1997.

Catucci, Stefano. *Per una filosofia povera: La Grande Guerra, l'esperienza, il senso: a partire da Lukács*. Torino: Bollati Boringhieri, 2003.

Chamayou, Grégoire. *Drone Theory*. Translated by Janet Lloyd. London: Penguin Books, 2015.

Crary, Jonathan. *Techniques of the Observer. On Vision and Modernity in the Nineteenth Century*. Cambridge, MA: MIT Press, 1991.

De Luna, Giovanni. *Il corpo del nemico ucciso: Violenza e morte nella guerra contemporanea*. Torino: Einaudi, 2006.

Didi-Huberman, Georges. *Quand les images prennent position: L'œil de l'histoire 1*. Paris: Les Éditions de Minuit, 2009.

Didi-Huberman, Georges. "Échantillonner le chaos. Aby Warburg et l'atlas photographique de la Grande Guerre." *Études photographiques*, no. 27 (18 November 2011). http://journals.openedition.org/etudesphotographiques/3173.

Didi-Huberman, Georges. *L'Album de l'art à l'époque du Musée Imaginaire de Malraux*. Paris: Hazan, 2013.

Fabi, Lucio. *La prima guerra mondiale 1915–1918*. Roma: Editori Riuniti, 1998.

Farocki, Harun. *Images of the World and the Inscription of War*. 16mm film, 75 mins (1988).

Foucault, Michel. "Of the Other Spaces: Utopias and Heterotopias." Translated by Jay Miskowiec. *Diacritics* 16, no. 1(1986): 22–27.

Friedrich, Ernst. *Krieg dem Kriege!*. Berlin: Freie Jugend Verlag, 1924.

Gibelli, Antonio. "Nefaste meraviglie: Grande Guerra e apoteosi della modernità." In *Storia d'Italia, Annali: Guerra e pace, 18*, 547–551. Torino: Einaudi, 2002.

Gioia, Paola, Marco Pizzo, and Adriano Santiemma. "Ricordando la prima guerra mondiale." *Digitalia. Rivista del digitale nei beni culturali*, no. 1(2012): 67–81.

Grossi, Erica. "La memoria nelle fotografie dalle trincee della Grande Guerra. La menomazione percettiva come testimonianza." In *The Great War in Italy: Representation and Interpretation*, edited by Patrizia Piredda, 196–205. Leicester: Troubador, 2013.

Grossi, Erica. "La violenza nelle sperimentazioni visuali del 'Petit Journal Illustré': Dispositivi, supporti, palinsesti (1890–1910)." In *Il lungo Ottocento e le sue immagini: Politica, media, spettacolo*, edited by Vinzia Fiorino, Gian Luca Fruci, and Alessio Petrizzo, 207–219. Pisa: ETS, 2013.

Guerri, Maurizio. *La mobilitazione globale: Tecnica, violenza, libertà in Ernst Jünger*. Udine: Mimesis, 2012.

Guillot, Hélène. "La section photographique de l'armée et la Grande Guerre: De la création en 1915 à la non-dissolution." *Revue historique des armées*, no. 258(2010): 210–217. http://journals.openedition.org/rha/6938.

Heartfield, John, and Kurt Tucholsky. *Deutschland, Deutschland über alles*. Hamburg: Rowohlt, 1929.

Hillman, James. *A Terrible Love of War*. New York: The Penguin Press, 2004.

Isnenghi, Mario. "La guerra rappresentata." *Rivista di storia e critica della fotografia*, no. 1 (1980).

Jünger, Ernst. *Das Antlitz des Weltkrieges: Fronterlebnisse deutscher Soldaten*. Berlin: Neufeld & Henius, 1930.

Jünger, Ernst, and Ferdinand Bucholz. *Der gefährliche Augenblick: Eine Sammlung von Bildern und Berichten*. Berlin: Junker und Dünnhaupt Verlag, 1931.

La Guerra Bianca. National Geographic Italia. 2016. Photographs by Stefano Torrione. www.stefanotorrione.com/portfolio_page/la-guerra-bianca/.

Langford, Martha. *Suspended Conversation: The Afterlife of Memory in Photographic Albums*. Montreal: McGill – Queen's University Press, 2001.

Le Corre, Florence. "Les albums de photographie: une lecture dirigée." In *Histoire & conservation d'un objet: Journée d'études du groupe photographique de la Section française de l'institut international de conservation*, edited by Musée national d'histoire naturelle (Paris, 26–27 novembre 1998), 19–25. Champs-sur-Marne: L'Indépendant, 2000.

Le Goff, Jacques. *La nouvelle histoire*. Paris: CEPL, 1978.

Leoni, Diego. *La guerra verticale: Uomini, animali e macchine sul fronte di montagna, 1915–1918*. Torino: Einaudi, 2015.

Mirzoeff, Nicholas. *Watching Babylon: The War in Iraq and Global Vision Culture*. New York: Routledge, 2005.

Mitchell, William J.T. *Picture Theory: Essays on Verbal and Visual Representation*. Chicago: The University of Chicago Press, 1995.

Mitchell, William J.T. *Cloning Terror: The War of Images, 9/11 to the Present*. Chicago: The University of Chicago Press, 2011.

Moholy-Nagy, László. *Malerei Fotografie Film*. München: Bauhaus Büchner no.8 – Albert Langen Verlag, 1925.

Mondini, Marco. *Alpini: Parole e immagini di un mito guerriero*. Roma-Bari: Laterza, 2008.

Mondzain, Marie-José. *Homo Spectator*. Paris: Bayard, 2007.

Patočka, Jan. *Heretical Essays in the Philosophy of History*. Translated by Erazim Kohák. Chicago: Open Court, 1996.

Pinotti, Andrea, and Antonio Somaini. *Cultura visual: Immagini, sguardi, media, dispositivi*. Torino: Einaudi, 2016.

Pizzo, Marco. "Fondo Fotografico della Prima Guerra Mondiale." In *Repertori del Museo Centrale del Risorgimento 1, Fotografie del Risorgimento Italiano*, 288–355. Roma: Gangemi Editore, 2004.

Quetin, Michel. "L'album photographique: un observatoire original indispensable de points de vue individuels sur le monde." In *Histoire & conservation d'un objet: Journée d'études du groupe photographique de la Section française de l'institut international de conservation*, edited by Musée national d'histoire naturelle (Paris, 26–27 novembre 1998), 5–52. Champs-sur-Marne: L'Indépendant, 2000.

Rouillé, André. "Éditorial: Vertige du nombre." *La Recherche Photographique – Collection*, no. 10(1991): 5–7.

Sebald, Winfried G. *On the Natural History of Destruction*. Translated by Anthea Bell. London: Hamish Hamilton, 2003.

Sicard, Monique. *La fabrique du regard: Images de sciences et appareils de vision, XVe–XXe siècle*. Paris: Odile Jacob, 1998.

Sontag, Susan. *Regarding the Pain of the Others*. New York: Farrar, Straus and Giroux, 2003.

Schwartz, Angelo. "La guerra rappresentata: La fotografia della Prima guerra mondiale in Italia. Usi, ruoli, funzioni." *Rivista di storia e critica della fotografia* 1, no. 1 (1980).

Traverso, Enzo. *L'histoire comme champ de bataille: Interpréter les violences du XXe siècle*. Paris: La Découverte, 2011.

Véray, Laurent. *La Grande Guerra au cinéma: de la gloire à la mémoire*. Paris: Éditions Ramsay, 2008.

Virilio, Paul. *War and Cinema: The Logistics of Perception*. Translated by Patrick Camiller. London: Verso, 1989.

Warburg, Aby. *Mnemosyne – Bilderatlas: Zur Ausstellung im Kunsthaus Hamburg*. Edited posthumously. Hamburg: Dölling und Galitz Verlag, 1929.

Warburg, Aby, Georg Thilenius, and Giulio Panconcelli-Calzia. *La Guerra del 1914: Rivista illustrata dei primi tre mesi, Agosto Settembre Ottobre*. Hamburg: Broschek & Co., 1914.

Warburg, Aby, Georg Thilenius, and Giulio Panconcelli-Calzia. *La Guerra del 1914–15: Rivista illustrata dei mesi Novembre Dicembre Gennaio Febbraio*. Hamburg: Broschek & Co., 1915.

Part III
Relationship between past, present, and future

8 Brothers – and sons – in arms

First World War memory, the life cycle, and generational shifts during the Second World War

Ashley Garber

The year 1939 sparked a collision of war times. The unfolding Second World War occupied a temporal space informed by and constructed in reference to previous conflicts, especially the First World War. British war planners adopted a conscription scheme to distribute manpower between military and civilian needs, in the hope of avoiding the large-scale military casualties and disruptions to war production suffered from 1914 to 1918.[1] A series of Neutrality Acts in the late 1930s sought to prevent a recurrence of America's 1917 intervention in European affairs.[2] Developing and implementing a shared Allied strategy required, among other things, balancing memories of the British Empire's failed amphibious assault at Gallipoli alongside suspicions that America might again withdraw from collective security for the post-war world.[3]

Scholars have recognized not only how the First World War shaped strategic and political approaches to the Second, but also how the Second World War altered the memory of the First. Work by Dan Todman, Mark Connelly, and Adrian Gregory has demonstrated, for example, how British interpretations of the Second World War solidified associations of 'futility' with the nation's efforts in the First.[4] Scholars of American memory, including Steven Trout and Mark Snell, have found that the United States' involvement in the First World War came to be dwarfed by its role in the Second.[5] Nevertheless, little is known about how this process of change unfolded from 1939 to 1945. Contemporaries in unpredictable circumstances mentally accommodated the contingent relationship between past, present, and future in the midst of uncertainty. Without the benefit of hindsight, they could not fully anticipate the ultimate significance or implications of the events in which their lives were entangled. Interpretive frameworks informed by the First World War had to be reconciled to Second World War developments in real time, requiring historical actors to renegotiate the meanings assigned to two competing yet related 'war times'. This study examines a select subset of the population for whom the memory of the Great War held fundamental significance: those who had fought in the conflict and still identified as veterans of that experience 20 years later.[6]

This chapter considers how members of the British and American Legions oriented themselves in relation to Second World War servicemen, tracing how the juxtaposition of these two 'war times' transformed memory narratives of the First World War. In the absence of scholarly consensus surrounding definitions, this essay draws on Dan Todman's conceptualisation of 'memory' as being "reference to an experienced event in the past".[7] As Jay Winter and others have demonstrated, 'memory' is inherently multi-vocal and develops over time.[8] This essay uses 'narratives' or 'discourses' whenever possible in order to acknowledge the multiple strands within memory discussions. Specific discourses of the Great War were leveraged after 1939 to construct a sense of comradeship in both Legions that could encompass veterans from two different generations. As recognised by Elwood Carlson, 'generations' represent "[s]hared experiences (usually with a temporal focus) [that] generate a group out of a collection of individual personalities", creating a potential source of lifelong social identity.[9] In contrast to age grades or cohorts, the term 'generation' allows for the consideration of historical influence on individual and group identities.[10] This work extends existing scholarship on generational history, which tends to focus on ideas of adolescent revolt and political mobilisation.[11] It also builds upon research by Antoine Prost, Gary Sheffield, Alistair Thomson, and others that places veterans' organisations at the centre of memory expression, formation, and rehearsal for former servicemen.[12] Organisational periodicals performed an important role in this process by providing a forum for discussion that spanned geographical barriers. This study focuses on periodicals released between 1939 and 1941 for the British Legion, and 1942 to 1944 for the American Legion.[13] As demonstrated in the analysis below, these time frames reflect the respective transition periods when First World War veterans in each group cognitively came to terms with a younger generation still largely absent from their membership ranks.

The Second World War prompted a re-evaluation, but not a complete abandonment, of the 'brotherhood of the trenches' camaraderie concept that initially anchored the British and American Legions. Recruiting Second World War veterans would extend the lifespan of each Legion while simultaneously deterring the formation of new, competing organisations. Incorporating ex-servicemen from a new global war into extant First World War groups, however, would require a definition of comradeship that could accommodate discrepancies in age, life phase, and war experience – in effect, a 'trans-generational' comradeship. To negotiate these divergences, discussions within the British and American Legions focused on connections that linked disparate war aims and service experiences of the two world wars as well as tangible familial bonds between the generations fighting them. The way in which conceptions of trans-generational comradeship developed in real time ultimately transformed First World War memory narratives. Asserting veteran relevancy and drawing ties with the contemporary conflict necessarily privileged certain memories of the First World War over others.

Britain and America experienced distinct world wars in both instances. In the First World War, Britain mobilised approximately half of its eligible male population.[14] Thirty-six per cent of men in uniform from Britain and its Empire became casualties of the war. In comparison, approximately 18 per cent of American male citizens enrolled for military duties,[15] and these men suffered a casualty rate of 8 per cent.[16] The war was also shorter for the United States, which officially entered into hostilities in April 1917 compared to Britain's entry in August 1914.[17] While the scale and duration of America's engagement with the war paled in comparison to its European associates, Jennifer Keene has contended that the salience and significance of the experience resonated deeply within American society via the conflict's ex-servicemen.[18] Similarly, each nation faced disparate Second World Wars. Again, the United States entered the fray years after Great Britain, declaring war in December 1941 following the surprise Japanese attack on Pearl Harbor. Ultimately, the United States mobilised 15 million people – one out of every nine citizens.[19] The country suffered 405,000 dead, comprising less than a third of 1 per cent of the total population at the time.[20] Britain raised a force of more than five million in the Second World War, with a considerable number of civilians also placed in the line of fire.[21] Developments in airplane technology and war tactics combined to leave much of Britain's population vulnerable to assaults by German *Luftwaffe* air forces. By the war's end in 1945, 1 per cent of the population had been killed, including 67,000 civilians.[22]

Different national experiences translated into distinct veterans' organisations, which emerged out of the First World War. Founded in 1921, the British Legion was an amalgamation of four voluntary associations created by ex-servicemen to preserve their wartime relationships and protect their interests in civilian life.[23] It claimed up to 11 per cent of Britain's Great War veterans as members at its height, with a peak of approximately 409,000 members in 1938.[24] Despite repeated efforts to influence government policies affecting ex-servicemen – particularly those of pension reform – scholars generally see the British Legion as having had a limited political impact.[25] Its presence was felt most strongly on the national level through its annual Poppy Day appeal, and locally through other charitable efforts and remembrance activities for the First World War.[26] Its American counterpart, founded in 1919, also played a central role in commemorations of the war.[27] In contrast to the British Legion, the American Legion exerted considerable influence in national and local politics, derived in part from its formidable size: the group reported a membership roll in excess of one million in 1931.[28] At any given point in the interwar period, it represented between 15 to 25 per cent of the former doughboys eligible for membership.[29] Additionally, the group's founding leaders were extremely well connected in the corridors of power, helping the organisation secure significant legislative victories in the 1920s and 1930s.[30]

The characterisations of the American Legion as a political player and the British Legion as a charity conventionally found in historiography reflect the divergent historical approaches to veteran support and status in each nation. By the First World War, there was a long-standing tradition in the United States of citizen-soldiers organising themselves upon discharge and campaigning for improved ex-service entitlements. They occupied a privileged position in society.[31] Such a precedent did not exist in Britain. In contrast, the history of

British veterans is one of neglect, both by the government they served and by their civilian counterparts. After the First World War, British ex-servicemen in need – especially the disabled – relied heavily on charitable aid to supplement the skeletal and inadequate system of state support.[32]

These differences aside, the two groups offer a fruitful opportunity for comparative analysis. Both Legions shared fundamental characteristics. The groups shared national leaderships dominated by the middle and upper classes, as well as hierarchical structures of local, regional, and national outposts.[33] While open to a range of veterans, organisational memberships tended towards white men from rural communities who championed traditional values.[34] Their conservative political agendas facilitated close working relationships with their respective states.[35] Most significantly for this study, both organisations drew heavily upon their Great War roots to define a sense of collective identity.[36] This war anchored group ideas about who counted as a veteran, his sociopolitical role and relationship with the state, and the consequences of armed conflict for society. This chapter explores how these conceptualisations changed in response to the Second World War and its generation, focusing on the subjects of shared war aims, service life, and the rhetoric of kinship.

The British Legion, 1939–1941

Even as the developing conflict came to outstrip its earlier counterpart, British Legion discussions identified similarities between the motivations for intervention in both world wars to help establish a preliminary basis for trans-generational comradeship. The new war was but a month old when British Legion author Frank H. Shaw rebranded the Great War as "the last war, as it will have to be called now, since this present conflict promises to be known to posterity as the Greatest War".[37] The *British Legion Journal* depicted Britain's decision for war in both cases as a defence of principles. In "another job of work in cleaning up Europe", British servicemen faced a "task to once more try to make the world a better and brighter place for all decent thinking people."[38] A regular contributor to the *Journal*, 'Kitchener's Man', wrote in December 1939 that "clearly complete links" existed between "this new generation" and "the 1914–1918 men", claiming that "For the first time in England's story have we had from one generation to another this complete succession of tradition in arms borne in the cause of human freedom".[39]

For this First World War veteran, the spirit of "arms borne in the cause of human freedom" offered the basis for trans-generational comradeship. Such a claim represented a revision to interwar interpretations of the First World War that had focused on peace. These British Legion narratives began shifting in the late 1930s, after repeated confrontations with Germany forced them to reconsider the sustainability of this assumed legacy.[40] Peace was conspicuously absent from the December 1939 article outlining the Great War's aims. Rather, according to 'Kitchener's Man', freedom and bettering the world had been the primary motivations all along.

The First World War served as a benchmark in Legion discussions throughout the so-called 'Phoney War' from September 1939 to May 1940. The perceived failure of the Treaty of Versailles and the German culpability for "again" "thrust[ing] Europe into such a hideous nightmare" constituted primary themes in an April 1940 article by former National Chairman Colonel G. R. Crosfield. Yet in other ways the connections drawn between the conflicts had begun to shift. In the same piece, Crosfield remarked: "We never dreamt that within 21 years we should have to face a sterner and tougher fight."[41] The comments reflected the organisation's – and the nation's – anticipations more than the realities at that point in the new war, but events soon confirmed these fears.[42] The fall of France forced the Legion to recognise that even Britain's First World War cemeteries on the continent had "now [been] desecrated by a war more hideous than that in which [the interred] gave their lives".[43] By the Annual Conference in May 1940, the unfolding conflict had shattered all illusions that the Great War had been waged and won for peace: "At Queen's Hall this year, the organised survivors of the 1914–18 shambles were compelled to accept defeat – not defeat of their ideals, but defeat of peace itself."[44] It is significant, however, that the *ideals* of the First World War generation remained undefeated. The memory of the conflict in Legion discourses continued to shift: whereas maintaining the peace had previously been emphasised, other principles for which Britain had fought now took centre stage: the defence of liberty and the progression of the modern state. This shift facilitated stronger ties between the First and Second World Wars. Rather than making the previous war irrelevant and futile, the Second World War effort was now a continuation of Britain's efforts in the First – even as it came to overshadow this predecessor.

That both generations of British servicemen fought against a similar enemy provided another continuity. Any distinction between Wilhelmine and Nazi Germany was given little notice by Legion contributors, one of whom referred to the country as "our late – and present – enemy".[45] Both regimes were seen to share a ruthless ambition for power. Just as "the veterans of the Great War [...] stood between German ambition and our national freedom", so too did the Second World War men now "stand again for civilisation against a reign of terror and lawless aggression".[46] Indeed, author 'Mons Star' believed that Britain's First World War dead "died to destroy exactly such a menace as this which now dominates civilisation".[47] Legion writers acknowledged the prowess of the German military machine in both conflicts, but asserted Britain's ability to triumph nonetheless: "Like the old timers of 14–18, the youngsters refuse to be overawed by the stamping and roaring of the German big shots."[48] German leaders were singled out for particular censure, blamed for disregarding the hard-won lessons of an earlier era. As Branch President Lloyd H. Fox expounded to West Buckland members, "The crime Hitler has committed against humanity becomes an immeasurably greater piece of villainy when we remember that he is an ex-Service man."[49] Brigadier-General T. N. S. M. Howard, President of Hampshire County, likewise drew a direct line from the German leaders of 1914 to those of 1939: "So far as we ex-Service men and women are concerned, Hitler and his group of despots still stand for the destruction of all that we fought for in the last war."[50] Continuing the same mission against the same enemy united two generations from two different wars into a shared sense of camaraderie.

Comparing aspects of service life between the two world wars brought out commonalities as well as disparities, and discussants used both to strengthen notions of trans-generational comradeship.[51] The 'old soldiers' reminisced in Legion newsletters about their escapades in uniform, retelling pleasant memories of romances and camaraderie as well as more bittersweet ones of homesickness.[52] These seemingly mundane experiences bonded men across military branches and outposts.[53] Still, in the comparisons between First and Second World War servicemen that peppered *Journal* articles, the latter were typically held up as superior. For example, National Chairman Major Sir Francis Fetherston-Godley reflected that "The present war soldier was a man who thought".[54] Identifying contemporary troops as 'thinking men' suggested, by contrast, that those of the First World War were not. Carried to its conclusion, the implication provided a subtle, but effective, means of bolstering wartime morale: if an unenlightened Tommy Atkins could win a war in 1918, then these 'thinking' men of 1939 certainly could do the same. Reiterating the advances of the present-day military over its predecessor worked to reassure veterans with serving sons that their children were in good hands – better hands, in fact, than they themselves had been. It fed into an ever-growing narrative that championed the Second World War generation as being superior to the First, minimising and, in many cases, simplifying the memory of what Britain had contributed and achieved in its previous conflict.

Current servicemen outstripped their fathers in other ways as well. The "air of businesslike efficiency about this outfit" struck 'Kitchener's Man' as a notable difference between the two armies. "There are no buttons to clean, no puttees to unravel – only a war to win and to win quickly", he reported. Cumbersome military regulations had been removed and these soldiers no longer stood on ceremony, which in the author's mind suggested a more egalitarian force. Maintaining the highly mechanised modern military required more skilled soldiers, lending an air of "equal importance and value" to all ranks. In the end, he judged the younger cohort to be "a worthy generation to follow in the steadfast footsteps of the older bearers of the flag of freedom" – these being the First World War veterans specifically of the British Legion.[55]

This focus on rank and equality reflected as well as reinforced disillusionment discourses about military leadership found in interwar cultural texts, previously explored at length by numerous scholars, most notably Paul Fussell.[56] The assertion that soldiers of this war respected one another across ranks evoked and reiterated wider memories of rank being an issue in the First World War. In stating that these equality-minded men would step readily into the British Legion, 'Kitchener's Man' was also affirming the egalitarian nature of the organisation itself. Officer eligibility for membership had been a point of contention between the constituent groups that joined together to form the Legion, and the role of officers continued to be debated in the following years.[57] The needs of the organisation thus contributed to how trans-generational comradeship was conceptualised. In this case, the need to uphold the principle of equality among Legion members, regardless of rank or class distinctions, influenced what was valued in Britain's Second World War military, and what was seen as lacking in that of the First.

Camaraderie between generations was further underscored through kinship frameworks that accommodated differences in age and life phase.[58] The fact that many First World War Legion members were sending their own sons and daughters to war in 1939 helped reframe life-cycle divergences as connections. As 'Mons Star' expressed it, "We have surrendered our sons to fight in this new crusade", calling it "a righteous one if ever one was begun."[59] "[T]he sons of Legionaries", Second World War servicemen were depicted as "carrying on the best traditions of their fathers".[60] Children attending Legion functions or winning medals in the forces earned mentions in branch reports to the *Journal.* [61] Death notices often reported the deceased Legion member's experience of the Great War as well as his children under military command in the current conflict.[62] One account from a First World War veteran who was again in uniform, training young recruits for the Second, suggested that these young men also recognised and found meaning in familial links: "If I indulge in reminiscences perhaps of Ypres and draw upon my experience there for an example, a shy well-spoken lad is sure to come to me after the lecture to say 'My father was at Ypres'."[63] Rather than distancing him from contemporaries, sharing personal memories of his experiences from 20 years ago tied the veteran trainer to his younger recruits by aligning him with familial role models in their lives, such as their fathers.

Legion members were encouraged to take on supportive – even 'parental' – roles for troops to enrich their successors' military experiences.[64] Programmes, such as sending comfort parcels and offering advice bureaux, sprang from the 'old soldier's' compassion for the new one, for "Their own troubles in the Great War period [gave] them a special sympathy with those who are going through the same trials now".[65] These practical efforts manifested the perceived connections in service experiences that linked the two generations. Indeed, the Legion asserted that "no one [understood the needs of serving men] better than men who served in the Great War", who had likewise endured the hardships of war.[66] Improving the conditions of service life was a way for elder veterans to assert the relevancy of their own experiences while simultaneously contributing to the current war in substantive ways.

Yet the man in uniform himself was not always the target of Legion concern. Discourses devoted much attention to service dependants. Caring for First World War widows and orphans had been a key component of comradeship ideals since the Legion's inception, and the organisation expanded these endeavours in the contemporary war.[67] Aiding Second World War dependants constituted a "direct service" that could be offered "by sympathetic Legionaries, all of whom ha[d] been through the same troubles themselves".[68] In addition to addressing a genuine need created by the departure of breadwinners to the forces – further exacerbated by aerial bombardments targeting civilians – the work extended comradeship practices to the younger generation, making them equals with Great War veterans.

The efforts also spoke to the memory of the First World War. The idea that ex-servicemen and their families had been forgotten or abandoned after the Armistice heightened the importance of providing for dependants, and this legacy added urgency to protecting the Second World War generation from a similar fate. Drawing attention to its record of providing for veteran families in the past, these discussions also highlighted the advantages that Legion comradeship offered to serving personnel. The need to develop a useful conception of trans-generational comradeship ended up privileging a First World War narrative that emphasised the lack of state support for service dependants.

The American Legion, 1942–1944

Upon the United States' entry into the Second World War, the American Legion expressed little doubt that this second conflict was a direct descendant of 'their' war. Like the British Legion, American Legion discussions focused on the ideals for which the Great War had been waged, not its outcomes, in order to validate their own experiences while simultaneously connecting them to those of the current generation. Incorporating Wilsonian language to explain the latest war linked it and its war aims intrinsically to the First World War.[69] Even still, it was recognised that this battle would outstrip the preceding one. The emerging generation may have been "embarking upon [a] similar adventure", but it was one "of far greater magnitude".[70] "[A] greater and fiercer World War," this "life-or-death struggle" would be "the toughest war we have ever fought".[71] Depicting contemporary servicemen as those "who follow[ed] in our foot-steps" in order "to help Uncle Sam win his toughest war" reaffirmed that First World War veterans had done important work in laying the groundwork for their successors.[72] With each generation "crusading" towards similar ends, men of both wars became comrades through a shared mission, and the lessons of the Great War became inherently relevant to the success of the unfolding battle.[73]

More so than their British counterparts, American Legion discourses positioned these shared ideals within a historic national tradition. As Chairman of the Legion's World War II Liaison Committee, J. Ernest Isherwood argued that America's defenders in 1943 fought "for the American ideal of 'life, liberty, and the pursuit of happiness,' [and] for 'government of the people, by the people, and for the people'" in the same way that "American boys have always fought". These references tied the latest war to pillars of American mythology such as the Revolutionary and Civil Wars.[74] Defending the ideals of the United States remained consistent across the ages, thereby making men of the Second World War comrades with all those who came before them.

Situating the war aims of both world wars within a longer tradition rallied support for the current war effort and also suppressed lingering doubts about the prior conflict. America's entry into the First World War was met with divided opinions among the citizenry in 1917, and such disputes continued to plague the war's legacy years later. Suspicions that commercial ties to the Allies had led the United States into the war discredited arguments that it had been waged to defend national security, and the failure to build a sustainable new world order in its aftermath tainted claims justifying the effort based on its outcomes.[75] Steven Trout has argued that American Legion narratives of memory gravitated towards the "'experiential' value" of the war during the interwar years in order to avoid addressing whether the war had been strategically or politically necessary.[76] Focusing on the ideals that reportedly underpinned both the First and Second World Wars reaffirmed the significance of the former experience to the present, regardless of the debates still surrounding it. Discussions about trans-generational comradeship in the Legion thus capitalised upon America's engagement in another conflict to solidify select understandings about the First World War.

American Legion discussants, like their British counterparts, also examined everyday service life for ways to connect the two wars and generations. For example, the importance of letters from home or card games with others in uniform helped to keep morale up in the 1942 military as much as in the 1917 one.[77] Caring for unit mascots, snapping photographs, and collecting souvenirs from battle zones occupied the time of serving men in both generations.[78] That being said, contemporaries recognised that the United States military of 1942–1945 was a very different force than its 1917–1918 counterpart. Soldiers were said to have more comfortable uniforms and packs, as well as washing bags designed to exterminate lice from clothing even in the front lines.[79] Improved facilities reportedly made for more comfortable camp life as well, and the troops enjoyed military rations of a superior quality, nutritional value, and taste than their predecessors.[80] Advances in mechanisation and engineering further distinguished the current forces.[81]

Interwar interpretations loomed large over discussions about these improved provisions. Jennifer Keene has argued that American combatants in the First World War believed that their military training had not prepared them for the fighting they encountered on the Western Front.[82] This view inflected later Legion discourses about that war. Stephen Trout has contended that interwar fiction in the *Magazine* often repeated a 'fools become heroes' narrative, in which "clumsy and incompetent" doughboys found unconventional means to surmount the odds in battle and courageously earn victories.[83] Legion political support for military preparedness measures had been rooted in this belief that the country had been caught ill-equipped for the First World War.[84] Examining the efficacy of contemporary forces thus stemmed from, and played into, this memory of poor preparation.

In addition to serving in a more advanced military than their First World War predecessors, Second World War men in uniform faced dissimilar living conditions in the lines, especially in the Pacific theatre. Associated Press war correspondent and former doughboy J. Norman Lodge prepared a comparative assessment of the New Georgia campaign in the Solomon Islands for the *Magazine* in January 1944. While the survivors of the "practice World War" complained about "such sissified things as deep dugouts and six-foot trenches", men fighting in the Pacific had it much worse, according to Lodge. Despite their filth, trenches at least allowed men to shelter alongside comrades while under fire, in contrast to the solitary confines of a jungle foxhole. The respite provided by deep dug-outs in the lines and civilian towns outside of them was also denied to South Pacific troops. Even the lack of lice was not a benefit there, because the men faced greater pests, including reptiles "using your trouser legs as garages". "You thought the Argonne was bad", he told his comrades, "and for 25 years you have been kidding the public that it was tough."[85] Although these points were intended to praise the men facing such appalling conditions rather than rebuke First World War veterans, such coverage of life in the Pacific served to displace associations of hardship with the Great War.

Thus – as in the British Legion case – American Legion comparisons of service life developed ideas of trans-generational comradeship while also positioning Second World War servicemen as superior to those of the First. Depicting the present-day military as advancing on the First World War experience created an intrinsic connection between the two. Although distinctions remained, a narrative of progress underpinned discussions of life in uniform for both generations. It also tied the conflicts together, building upon existing claims that the United States had been unprepared to face war in 1917. The American GI of the Second World War was depicted as a better-equipped and better-cared-for soldier, fighting in more horrific conditions than his doughboy predecessor. Portraying military life in this way simplified the legacy of the First World War, privileging certain interpretations of it as a result.

Former doughboys could also be literally joined to the Second World War, and discourses accentuated these bonds. American Legion discussions highlighted experiences of First World War ex-servicemen who were again donning military uniforms in the Second. The *Magazine* used the term "retread" for these men, drawing upon a metaphor of old "tires" being retrofitted with updated treads.[86] It is striking that these men received more attention in American discourses than in British ones, especially given that more First World War veterans were eligible to wear military uniforms again in Britain than in the United States due to service-age regulations.[87] Articles detailing how these older men connected with their younger comrades strengthened ideas of trans-generational comradeship.[88] In these cases, trans-generational comradeship was more than a constructed concept based on similarities in service experiences. It sprang from the traditional understanding of comradeship rooted in training or fighting alongside one another. 'Retreads' literally brought the ex-service and serving worlds together, not just through camaraderie in the ranks, but also by organising American Legion Posts and meetings at military sites.[89] Their unique position enabled them to transcend the delineations of active duty and home front, creating a bridge for fostering trans-generational comradeship into life after discharge as well.

Contributors to the *Magazine* often depicted "retreads" as the lucky ones of their generation. When General Pershing offered his "heartiest congratulations" to his "comrades who [were] again serving with the armed forces" in April 1942, he echoed sentiments expressed by other First World War veterans.[90] Even men destined by age or health for short stints in uniform this time around were "lucky [...] to live over [their] 1917–18 experience with the '42 edition".[91] Having such "good fortune" to serve their country twice over distinguished them from their peers, who were often depicted as wishing they were also wearing "retread" shoes.[92] Such descriptions certainly satisfied propaganda purposes, reiterating that wartime service to the state was a desirable privilege of citizenship in a democracy. In addition to these messages about the current war, though, accounts by or of "retreads" in the *Magazine* implicitly extolled the First World War as well. After all, the last war could not have been entirely awful or futile if those who were serving again were identified as "lucky" and "fortunate".

Kinship links also proved powerful in constructing trans-generational comradeship within American Legion discussions, just as they did in the British case. Yet juxtaposing the older veteran 'parent' with the younger 'child' in uniform revealed underlying anxieties about divergent life phases and their implications. The beginning of a new world war, for which many were too old to serve, combined with the 25th anniversary of their involvement in the previous war, forced First World War veterans to confront the realities of aging.[93] "We'll have to start admitting that we're old vets", John J. Noll told his readers.[94] Inviting serving men to join their ranks reiterated this sense that older veterans were past their prime. The "young fellows and young women who are now in uniform" were "replacing" their predecessors who were "oldtimers [sic] who couldn't pass a physical even for a job as latrine orderly or captain of the head".[95] In turn, they were likewise invited to replace Great War veterans in the Legion upon discharge.

Examples of familial links between the generations depicted aging as a disruptive process, particularly in wartime. For example, Legion member W. L. Paul, Sr., submitted a photograph to the *Magazine* of himself wearing olive drab and standing with his toddler son in 1918, alongside a present-day one in which the son wore the uniform instead. Behind its humour, his letter describing the photos struck a tone of apprehension. "Don't get the mistaken idea from comparison of the snaps that this old veteran has shrunken in height", he warned readers, explaining: "I still stand 5 feet 10 inches, but that son of mine towers to 6 feet 4 inches, plus!" The obvious pride in his son did not diminish an awareness of his own aging and his fear that what the camera had captured would be taken as evidence of physical decline. Both photographs showed "the civilian [...] wearing the Army cap", playfully highlighting the transition of the son from civilian to combatant status while his father moved in reverse.[96] Advancing through different life phases had made the son the protector of his aged father – a reversal of contemporary understandings that associated fatherhood with safeguarding the family.[97]

Watching their children go to war in their stead also prompted some veterans to reassess what they had achieved in their younger days. "Twenty-five years ago today", Legion member C. A. Barrett had "joined the Army to fight a war / To rid the earth of a festering sore". Having failed to accomplish this task, Barrett lamented that "now my son has gone to war / To finally finish an unfinished chore". While acknowledging the military victory, Barrett's poem portrayed the American First World War effort as an ultimate failure, because it did not eradicate the underlying problem. His generation was now accountable for ensuring "that no father may need agin [sic] note, / 'I've sent my son to carry a gun / To finish the job I should have done'". Victory in this new war would be as hollow as that of the last unless "the sons of our sons may live in peace / And the fear of war on earth may cease".[98] In Barratt's view, the failures of the post-war world had trumped any military successes. The fact that his son had to bear arms in another conflict offered a striking rebuke of his own generation. Rather than securing a more positive future for their offspring, Barratt believed that former doughboys had condemned their children to their same fate. The conflict of 1917–1918 is remembered here as a failed victory, and correcting those mistakes represented the ultimate expression of comradeship from men of that war to men fighting in the new one.

That the meaning of their efforts might be undermined in the future also connected ex-servicemen of both world wars. When Virginia banks announced a cancellation of Armistice Day observances in order to support the new war effort, John J. Wicker interpreted these temporary measures as indications that the First World War was being lost to history, due in part to aging. His poetic reply sarcastically commented, "Armistice Day? Let's all forget" because those who fought were "Old, disabled, dead – you know / We cannot use them now!" While proponents of the measure reasoned, "Oh yes, it's true, with lavish pride / We cheered them – waved goodbye! / But that was in their useful days – / Now let their memory die!", Wicker's conclusion reasserted the importance of remembering the earlier war, declaring "They'll never die!" Interestingly, the poem was inspired not by Wicker's own fears of redundancy, but those of the next generation. He reportedly overheard a current seaman remark, "Well, I suppose in twenty years or so everybody will be forgetting about us too".[99] Wicker's poem shows an awareness that veterans of any war had a limited time of usefulness in society's perceptions – and that this risk of being 'forgotten' ultimately united them.

Concern for the home they were defending further joined troops across wars. American Legion members echoed British Legion ones in believing that they possessed a unique understanding of the needs and concerns of servicemen, and they sought to leverage these insights through support programmes, which became an important wartime expression of comradeship.[100] Yet while British Legion veterans focused on protecting dependants for those fighting, the American Legion efforts centred on protecting the "American way of life".[101] "They want the America they are fighting for to be here when they return", J. Ernest Isherwood said of serving personnel in September 1943.[102] National Commander Warren H. Atherton warned that Legion members "on the home front must be alert to preserve that for which [those on the front lines] fight".[103] The hopes for America's future, cultivated through defending the nation, transcended generations and wars to unite all those who had worn the uniform, thus making the preservation of that future an act of trans-generational comradeship.

'Protecting the home' had distinct meanings for the British and American Legions, corresponding to divergent experiences on the home front in each country. In addition to the *Luftwaffe*, British civilians faced food and material shortages as well as inflation.[104] Rationing became "the most potent symbol" of the restrictions under which civilians lived during the Second World War.[105] Civilian consumption decreased by 22 per cent during the war. By contrast, the United States saw an increase in civilian consumption. Families that had struggled with the burdens of unemployment or financial insecurity during the Great Depression now encountered an industrial job market in which demand outstripped supply. Although directed primarily towards war production, the industrial revitalisation benefited consumers as well.[106] In David Kennedy's words, "Most Americans had never had it so good" as during the Second World War.[107] These marked variations between wartime civilian life in each nation help explain why the American Legion could afford to be less concerned than their British counterparts about the families left behind by serving personnel.

Moreover, the groups' differing emphases reflected interpretations of the First World War specific to each national context. As discussed earlier, the British Legion fretted that the nation would betray the sacrifices of servicemen by neglecting their dependants. The American Legion, however, was more concerned that society would abandon the principles for which men in uniform were fighting. Demands from American Legion members to "win the war; win the peace, and return to the normal American way of life" mirrored narratives that America had 'lost the peace' after the Armistice.[108] In the aftermath of the military victory, 'subversive' ideologies such as communism and fascism had taken hold and, in the Legion's view, threatened the ideals for which they had fought.[109] Legion efforts to preserve American ideals at home for the serving generation abroad thus reinforced the idea that this had not been achieved for them.

Conclusion

The outbreak of the Second World War prompted a collision of war times that proved especially problematic for First World War veterans' associations with collective identities rooted in the prior conflict. In the midst of shifting circumstances, the British and American Legions faced the possibility that millions of younger men with differing service experiencces would seek entry into their membership ranks. Coming to terms with the present required confronting the past, but the parameters for this confrontation were defined by the needs of the present moment. Those conditions were in constant flux – unstable and unpredictable to those living through them. Renegotiating the past in light of the present was thus an ongoing process. The memory of the First World War was transformed through real-time cognitive adjustments made in part to develop conceptions of trans-generational comradeship that could anchor the collective identity of both groups moving forward.

Distinctive national contexts nuanced discourses in particular ways, but the shared approaches to fostering this trans-generational comradeship are striking. Discussions in both Legions looked to unite disparate cohorts by identifying similarities in war aims and service life. Knitting differences into a cumulative narrative of progress – wherein one set of servicemen built upon the experiences of the other – accommodated these divergences while also constructing underlying connections. Likewise, utilising the rhetoric of kinship reinforced existing ties between the cohorts while allowing for disparate ages and life phases. These approaches not only joined ex-servicemen from different generations and conflicts, but they also confirmed the pertinence of older veterans' experiences for their younger counterparts. The specific ways in which developing ideas of trans-generational comradeship affected narratives of the First World War in each group reflected extant interpretations unique to each national and organisational context. Yet the commonalities of the process itself speak to its universal character, demonstrating that organised First World War ex-servicemen across varied contexts employed similar discursive techniques to adapt existing interpretive frameworks in response to the unfolding Second World War.

Concepts of 'generations' served important functions in constructing the collective identities and memory discourses of the British and American Legions. Existing generational histories concentrate upon a single cohort's perspective, ignoring that 'generations' are defined in relation to one another.[110] It is the interactions with other historically situated cohorts that create the boundaries of generational identity.[111] Incorporating the relational dynamic of generations into historical studies extends understandings of identity formation and its salience in specific contexts. At its root, memory is also relational. In the Legions, recognising a new generation required contemporaries to reflect upon their own experiences and the events of their formative years, as well as how these differed from later circumstances. Such reflections shaped the characteristics attributed to the emerging cohort and the meaning of preceding historical events. Looking at two generations and how they accommodated each other illuminates the real-time evolution of historical memory.

Trans-generational comradeship remained a crucial concern and challenge for both groups as they moved beyond the Second World War. The First World War generation's obligation to assist its Second World War comrades with the transition back to civilian life in peacetime affected the discussions and actions of both groups in the realm of post-war planning.[112] While the engagement of Second World War ex-servicemen with these concepts of trans-generational comradeship has yet to be studied in detail, the influx of new members following demobilisation suggests that younger veterans did not necessarily feel alienated by the Legions' Great War roots. By August 1946, for example, the American Legion could boast approximately 1,651,322 members from the Second World War generation.[113] Graham Wootton estimated that, of the more than one million members claimed by the British Legion in 1956, 60 per cent had served in the Second World War.[114] These men would come to dominate the ranks as well as leadership circles, inheriting ownership for the groups from their Great War predecessors. In turn, they would also face the difficulties of incorporating new generations into their organisations, continuing the processes of reconciling multiple war times and memory narratives via trans-generational comradeship in succeeding decades.

Notes

1 Calder, *The People's War*, 51; Marble, "Filling the Ranks", 585–590.
2 Doenecke and Wilz, *From Isolation to War*, 71–73, 78–79, 82, 88, 115.
3 Kennedy, *Freedom from Fear*, 576, 599, 697, 701, 679–680, 854.
4 Todman, *The Great War: Myth and Memory*, 94; Connelly, *The Great War, Memory and Ritual*, 8–9; Gregory, *The Last Great War*, 4–5, 273–278, 293–296.
5 Trout, *On the Battlefield of Memory*, esp. 222–247; Snell, *Unknown Soldiers*, esp. 16–37.
6 The terms 'ex-servicemen' and 'veteran' are used to describe former members of all military branches, including those in combatant and non-combatant roles. Likewise, terms such as 'serving personnel', 'servicemen', or 'troops' represent those in uniform across the service branches during the Second World War. The army perspective, however, dominates discourses in both organisations, so monikers traditionally associated with this specific branch, such as 'doughboy', 'old soldier', and 'Tommy Atkins', have been used more broadly in this piece. Discussions are also notably coded with a male voice, although female auxiliary associations did exist. Wootton, *The Official History of the British Legion*, 36, 75–6, 104, 243–252; Barr, *The Lion and the Poppy*, 48–52; Pencak, *For God and Country: The American Legion, 1919–1941*, 297–301; Trout, *On the Battlefield of Memory*, 100–105.
7 Winter, *Remembering War*, 135–136, 198.
8 *The Great War: Myth and Memory, 26.*
9 Carlson, "Generations as Demographic Category", 15.
10 Carlson, "Generations as Demographic Category", 15–16. While age grades lay at the intersection of aging and prescribed social roles, 'cohort' "stands for a set of individuals who pass some crucial life stage at approximately the same time, like marriage, first employment, and especially birth". Carlsson and Karlsson, "Age, Cohorts and the Generation of Generations", 710.
11 Examples of scholarship touching on these two themes across different contexts include: Wohl, *The Generation of 1914*; Klatch, *A Generation Divided*; Fulbrook, *Dissonant Lives*; Foster, *Vivid Faces: The Revolutionary Generation in Ireland 1890–1923.*

12 Prost, *In the Wake of War*, 19–23; Sheffield, *Leadership in the Trenches*, 131; Thomson, *ANZAC Memories*, 7–11, 213–215; Todman, *The Great War: Myth and Memory*, 187–217.

13 A range of authors contributed to the *British Legion Journal* (*BLJ*) and *American Legion Magazine* (*ALM*), and it was not uncommon for pieces to be printed without attribution. For consistency, the sources are identified by periodical, rather than author.

14 Winter, *The Great War and the British People*, 28.

15 This percentage was calculated by the author. Over 23 million American men between the ages of 18 and 45 were registered under the Selective Service Act of 1917. The United States' forces ultimately mustered 4.4 million in the war. Chambers, "Conscription", 180–182; Keene, "North America", 518.

16 Winter, "Demography", 249.

17 Logistical challenges further delayed American troops from reaching combat zones, even after war had been declared. Keene, "North America", 519.

18 Keene, *Doughboys, the Great War, and the Remaking of America*, ix–x, 7, 214.

19 Kennedy, *Freedom from Fear*, 747.

20 Bessel, "Death and Survival in the Second World War", 262.

21 Field, "Civilians in Uniform", 121.

22 Bessel, "Death and Survival in the Second World War", 262.

23 Barr, *The Lion and the Poppy*, 9, 11–21; Wootton, *The Official History of the British Legion*, 1–27.

24 Barr, *The Lion and the Poppy*, 2, 57; Wootton, *The Official History of the British Legion*, 305.

25 Cohen, *The War Come Home*, 17, 19, 52–55, 59; Barr, *The Lion and the Poppy*, esp. 119–149.

26 Gregory, *The Silence of Memory*, esp. 51–92.

27 Trout, *On the Battlefield of Memory*, 47, 62; Pencak, *For God and Country: The American Legion, 1919–1941*, 48, 287–288; Piehler, *Remembering War the American Way*, 114, 121.

28 Pencak, *For God and Country: The American Legion, 1919–1941*, xii–xiii, 107–143; Rumer, *The American Legion: An Official History, 1919–1989*, 23.

29 Pencak, *For God and Country: The American Legion, 1919–1941*, 82.

30 Ortiz, *Beyond the Bonus March and GI Bill*, 17; Pencak, *For God and Country: The American Legion, 1919–1941*, 35, 49, 52–54, 62–64, 80, 91; Rumer, *The American Legion: An Official History, 1919–1989*, 12, 104–109. Legion members called these men the 'kingmakers'. The list included men such as Theodore Roosevelt, Jr., (son of the former US president), Bennett Champ Clark (son of the former Speaker of the House and future congressman himself), and William 'Wild Bill' Donovan (a distinguished AEF soldier who went on to lead the Office of Strategic Services in the Second World War). The group's lobbying efforts were instrumental in the creation of the Veterans Administration after the First World War and the securing the so-called 'GI Bill of Rights' for the Second World War generation. The Legion was also involved to a degree in the interwar fight for adjusted compensation – the 'Bonus' – but leadership was slow to endorse the measure. Ortiz, *Beyond the Bonus March and GI Bill*, 4, 32–65, 187–206; Keene, *Doughboys, the Great War, and the Remaking of America*, 184–187, 205–214; Pencak, *For God and Country: The American Legion, 1919–1941*, 170–207, 319; Rumer, *The American Legion: An Official History, 1919–1989*, 187–194, 216–223, 243–248.

31 Ross, *Preparing for Ulysses*, 2; McConnell, *Glorious Contentment*, 123–165; Mettler, "Foreword", xi–xii; Pencak, *For God and Country: The American Legion, 1919–1941*, 6, 25, 31.

32 Cohen, *The War Come Home*, esp. 19–20, 26–46; Gregory, *The Silence of Memory*, 93–95; Gregory, *The Last Great War*, 266.

33 Barr, *The Lion and the Poppy*, 35–34, 68–74, 25–32; Pencak, *For God and Country: The American Legion, 1919–1941*, 49–54, 89–103.
34 Barr, *The Lion and the Poppy*, 48–52, 58–66, 191–195; Pencak, *For God and Country: The American Legion, 1919–194*, 81–83; Ortiz, *Beyond the Bonus March and GI Bill*, 19–20. This dominance of white male members reflected the general composition of First World War forces in the respective countries, although African American veterans were proportionately underrepresented in the American case. Pencak, *For God and Country: The American Legion, 1919–1941*, 68–69; Keene, *Doughboys, the Great War, and the Remaking of America*, 177–178; Trout, *On the Battlefield of Memory*. The American Legion's members were predominantly middle-class whereas the ranks of the British Legion came mainly from working-class backgrounds. Despite these class differences, each group's service-oriented activities and opposition to radical social or political change suggests members held conservative outlooks in common. Barr, *The Lion and the Poppy*, 64–67, 86, 193–195; Pencak, *For God and Country: The American Legion, 1919–1941*, 5–7, 42–47, 81, 104, 280–295, 320–321.
35 Cohen, *The War Come Home*, 52–57; Ortiz, *Beyond the Bonus March and GI Bill*, esp. 7, 11–12, 32–65.
36 Cohen, *The War Come Home*, 17–19, 52; Barr, *The Lion and the Poppy*, 64–67; Pencak, *For God and Country: The American Legion, 1919–1941*, 48; Trout, *On the Battlefield of Memory*, 47, 62.
37 *BLJ* 19:5, 1939, 147. See also Gregory, *The Last Great War*, 5.
38 *BLJ* 19:6, 1939, 176, 174.
39 *BLJ* 19:6, 1939, 176.
40 Barr, *The Lion and the Poppy*, 151–190; Wootton, *The Official History of the British Legion*, 252–254.
41 *BLJ* 19:10, 1940, 283.
42 Grayzel, *At Home and Under Fire*, 251–294; Calder, *The People's War: Britain 1939–1945*, 35–76.
43 *BLJ* 20:1, 1940, 6.
44 *BLJ* 19:12, 1940, 336.
45 *BLJ* 19:8, 1940, 233.
46 *BLJ* 19:5, 1939, 140; *BLJ* 19:6, 1939, 177.
47 *BLJ* 19:5, 1939, 140.
48 *BLJ* 19:6, 1939, 176. See also *BLJ* 20:5, 1940, 122.
49 *BLJ* 19:8, 1940, 247.
50 *BLJ* 19:6, 1939, 177.
51 For details on the military experience of citizen-soldiers in the First World War, see: McCartney, *Citizen Soldiers*; Keene, *Doughboys, the Great War, and the Remaking of America*.
52 *BLJ* 19:6, 1939, 176; *BLJ* 19:4, 1939, 130.
53 *BLJ* 19:4, 1939, 127.
54 *BLJ* 19:9, 1940, 260.
55 *BLJ* 19:6, 1939, 176.
56 Fussell, *The Great War and Modern Memory*, esp. pp. 90–98; Sheffield, *Forgotten Victory*, esp. 5–8; Winter, *Sites of Memory, Sites of Mourning*, esp. 186–196; Todman, *The Great War: Myth and Memory*, esp. 73–120.
57 Wootton, *The Official History of the British Legion*, 2, 12, 95–99.
58 The average age of British Legion members in 1937 was reportedly 48 (British Legion Report of the Special Committee, 1937, 9).
59 *BLJ* 19:5, 1939, 140.
60 *BLJ* 19:12, 1940, 336, 335.
61 *BLJ* 20:7, 1941, 193; *BLJ* 20:9, 1941, 245; *BLJ* 20:2, 1940, 50.
62 *BLJ* 20:2, 1940, 51; *BLJ* 20:8, 1941, 217; *BLJ* 20:3, 1940, 83.

63 *BLJ* 19:6, 1939, 176.
64 *BLJ* 19:7, 1940, 207; *BLJ* 20:9, 1941, 246–247.
65 *BLJ* 20:9, 1941, 229. Examples of programmes include: *BLJ* 19:6, 1939, 174; *BLJ* 19:5, 1939, 160.
66 *BLJ* 20:6, 1940, 144.
67 *BLJ* 20:9, 1941, 229, 244; *BLJ* 19:5, 1939, 160, 164; *BLJ* 20:4, 1940, 103; *BLJ* 20:8, 1941, 223; *BLJ* 19:9, 1940, 252.
68 *BLJ* 20:3, 1940, 60.
69 *ALM* 32:1, 1942, 1, 23; *ALM* 33:5, 1942, 1; *ALM* 34:1, 1943, 31.
70 *ALM* 32:4, 1942, 34.
71 *ALM* 33:6, 1942, 46; *ALM* 34:4, 1943, 22; *ALM* 32:4, 1942, 8.
72 *ALM* 33:2, 1942, 32; *ALM* 34:2, 1943, 27.
73 *ALM* 34:1, 1943, 31.
74 *ALM* 35:3, 1943, 21.
75 Piehler, *Remembering War the American Way*, 114–116.
76 Trout, *On the Battlefield of Memory*, 69, 71.
77 *ALM* 35:3, 1943, 20; *ALM* 36:4, 1944, 48.
78 *ALM* 35:5, 1943, 15, 16; *ALM* 35:6, 1943, 38.
79 *ALM* 34:1, 1943, 28; *ALM* 35:6, 1943, 11; *ALM* 37:6, 1944, 49.
80 *ALM* 32:5, 1942, 32–33; *ALM* 33:4, 1942, 16, 17, 60, 62.
81 *ALM* 32:5, 1942, 19; *ALM* 33:6, 1942, 58. These claims were generally true. Kennedy, *Freedom from Fear*, 709–716.
82 Keene, *Doughboys, the Great War, and the Remaking of America*, 43.
83 Trout, *On the Battlefield of Memory*, 82, 65, 84.
84 *ALM* 24:1, 1938, 11.
85 *ALM* 36:1, 1944, 10, 11, 44, 46.
86 *ALM* 35:6, 1943, 10.
87 Griffin, "Veterans at War: The American Legion and Civilian Mobilization in World War II", 40.
88 *ALM* 35:6, 1943, 10; *ALM* 36:2, 1944, 16.
89 *ALM* 34:4, 1943, 44; *ALM* 34:6, 1943, 29; *ALM* 35:4, 1943, 25.
90 *ALM* 32:4, 1942, 11.
91 *ALM* 32:5, 1942, 52.
92 *ALM* 36:2, 1944, 16; *ALM* 36:3, 1944, 25.
93 The average age of American Legion members in 1938 was reportedly 44 to 45. *ALM* 24:2, 1938, 27.
94 *ALM* 32:4, 1942, 34.
95 *ALM* 35:5, 1943, 14.
96 *ALM* 36:5, 1944, 24.
97 For more on fatherhood at this time, see LaRossa, *Of War and Men*, 51–63.
98 *ALM* 36:5, 1944, 9.
99 *ALM* 34:1, 1943, 2, 54, 55.
100 *ALM* 33:2, 1942, 14, 15; *ALM* 35:3, 1943, 21; *ALM* 32:4, 1942, 59–60; *ALM* 35:2, 1943, 25; *ALM* 34:2, 1943, 29; *ALM* 33:4, 1942, 54.
101 A notable example of discussions focusing on dependant care was *ALM* 35:3, 1943, 21, 43.
102 *ALM* 35:2, 1943, 44.
103 *ALM* 37:3, 1944, 6.
104 Todman, *Britain's War*, 607–609.
105 Connelly, *We Can Take It!*, 157; Todman, *Britain's War*, 607–609.
106 Kennedy, *Freedom from Fear*, 644–647.
107 Kennedy, *Freedom from Fear*, 646.

108 *ALM* 34:2, 1943, 26; "Summary of Proceedings, Annual National Convention of 1942", American Legion National Headquarters, Indianapolis, 33. See also Piehler, *Remembering War the American Way*, 114–115.

109 *ALM* 26:4, 1939, 18; *ALM* 25:2, 1938, 59; *ALM* 24:5, 1938, 35.

110 Carlson, "Generations as Demographic Category", 24; Berghoff, Jense, Lubinski, and Weisbrod, "Introduction", 7.

111 Carlson, "Generations as Demographic Category", 25; Jaeger, "Generations in History", 274.

112 This was particularly true for debates over state entitlements or benefits. These themes are explored in greater detail in my forthcoming dissertation: Garber, "Toward a 'Two-War Legion': Renegotiating First World War Memory in the British and American Legions, 1938-1946".

113 Rumer, *The American Legion*, 283.

114 Wootton, *The Official History of the British Legion*, viii.

Bibliography

Printed primary sources

American Legion Magazine, 1938, 1942–1944. Indianapolis: American Legion National Headquarters.

British Legion Journal, 1939–1942. London: The Royal British Legion National District Offices.

"British Legion Report of the Special Committee established for the purpose of investigating the problem of Prematurely Aged Ex-Service Men." London: The Royal British Legion National District Offices, 1937.

"Summary of Proceedings, Annual National Convention of 1942."Indianapolis: American Legion National Headquarters.

Printed secondary sources

Barr, Niall. *The Lion and the Poppy: British Veterans, Politics, and Society, 1921–1939*. Westport: Praeger, 2005.

Berghoff, Hartmut, Uffa Jensen, Christina Lubinski, and Bernd Weisbrod. "Introduction." In *History by Generations: Generational Dynamics in Modern History*, edited by Hartmut Berghoff, Uffa Jense, Christina Lubinski, and Bernd Weisbrod. Göttengen: Wallstein Verlag, 2013.

Bessel, Richard. "Death and Survival in the Second World War." In *The Cambridge History of the Second World War, Volume 3: Economy, Society and Culture*, edited by Michael Geyer and Adam Tooze, 252–276. Cambridge: Cambridge University Press, 2015.

Calder, Angus. *The People's War: Britain 1939–1945*. London, 1992 [orig. 1969].

Carlson, Elwood. "Generations as Demographic Category: Twentieth-Century U.S. Generations." In *History by Generations: Generational Dynamics in Modern History*, edited by Hartmut Berghoff, Uffa Jense, Christina Lubinski, and Bernd Weisbrod, 15–38. Göttengen: Wallstein Verlag, 2013.

Carlsson, Gosta, and Katarina Karlsson. "Age, Cohorts and the Generation of Generations." *American Sociological Review* 35, no. 4(1970): 710–718.

Chambers, J. Whiteclay II. "Conscription." In *The Oxford Companion to American Military History*, edited by John Whiteclay Chambers II, 180–182. Oxford: Oxford University Press, 2000.

Cohen, Deborah. *The War Come Home: Disabled Veterans in Britain and Germany, 1914–1939*. Berkeley: University of California Press, 2001.

Connelly, Mark. *The Great War, Memory and Ritual: Commemoration in the City and East London, 1916–1939*. Bury St Edmunds: The Boydell Press, 2002.

Connelly, Mark. *We Can Take It! Britain and the Memory of the Second World War*. Harlow: Pearson Longman, 2004.

Doenecke, Justus D. and John E. Wilz. *From Isolation to War: 1931–1941*. 4th edn. Chichester: Wiley Blackwell, 2015.

Field, Geoffrey. "'Civilians in Uniform:' Class and Politics in the British Armed Forces, 1939–1945." *International Labor and Working Class History* 80 (Fall 2011): 121–147.

Foster, Roy F. *Vivid Faces: The Revolutionary Generation in Ireland 1890–1923*. London: Penguin, 2014.

Fulbrook, Mary. *Dissonant Lives: Generations and Violence through the German Dictatorships*. Oxford: Oxford University Press, 2011.

Fussell, Paul. *The Great War and Modern Memory*. Oxford: Oxford University Press, 2013 [orig. 1975].

Garber, Ashley. "Toward a 'Two-War Legion': Renegotiating First World War Memory in the British and American Legions, 1938-1946." DPhil dissertation, Oxford University, 2018.

Grayzel, Susan R. *At Home and Under Fire: Air Raids and Culture in Britain from the Great War to the Blitz*. Cambridge: Cambridge University Press, 2012.

Gregory, Adrian. *The Silence of Memory: Armistice Day 1919–1946*. Oxford: Berg, 1994.

Gregory, Adrian. *The Last Great War: British Society and the First World War*. Cambridge: Cambridge University Press, 2008.

Griffin, Christopher W. "Veterans at War: The American Legion and Civilian Mobilization in World War II." PhD dissertation, Florida State University, 2012.

Jaeger, Hans. "Generations in History: Reflections on a Controversial Concept." *History andTheory* 24, no. 3(1985): 273–292.

Keene, Jennifer D. *Doughboys, the Great War, and the Remaking of America*. Baltimore: John Hopkins University Press, 2001.

Keene, Jennifer D. "North America." In *The Cambridge History of the First World War: Volume 1: Global War, The Cambridge History of the First World War*, edited by Jay Winter, 511–532. Cambridge: Cambridge University Press, 2014.

Kennedy, David M. *Freedom From Fear: The American People in Depression and War, 1929–1945*. Oxford: Oxford University Press, 1999.

Klatch, Rebecca E. *A Generation Divided: The New Left, the New Right, and the 1960s*. Berkeley: University of California Press, 1999.

LaRossa, Ralph. *Of War and Men: World War II in the Lives of Fathers and their Families*. Chicago: University of Chicago Press, 2011.

Marble, Sanders. "Filling the Ranks: Conscription and Personnel Policies." In *The Cambridge History of the Second World War, Volume 1: Fighting the War*, edited by John Ferris and Evan Mawdsley, 585–607. Cambridge: Cambridge University Press, 2015.

McCartney, Helen. *Citizen Soldiers: The Liverpool Territorials in the First World War*. Cambridge: Cambridge University Press, 2005.

McConnell, Stuart. *Glorious Contentment: The Grand Army of the Republic, 1985–1900*. Chapel Hill: The University of North Carolina Press, 1992.

Mettler, Suzanne. "Foreword." In *Veterans' Policies, Veterans' Politics: New Perspectives on Veterans in the Modern United States*, edited by Stephen R. Ortiz, xi–xiv. Gainesville: University Press of Florida, 2012.

Ortiz, Stephen R. *Beyond the Bonus March and GI Bill: How Veteran Politics Shaped the New Deal Era*. New York: New York University Press, 2010.

Pencak, William. *For God and Country: The American Legion, 1919–1941*. Boston: Northeastern University Press, 1989.

Piehler, G. Kurt. *Remembering War the American Way*. Washington: Smithsonian Institution Press, 1995.

Prost, Antoine. *In the Wake of War: 'Les Anciens Combattants' and French Society*. Translated by H. McPhail. The Legacy of the Great War Series, edited by Jay Winter. Providence: Berg, 1992.

Ross, David R. B. *Preparing for Ulysses: Politics and Veterans during World War II*. New York: Columbia University Press, 1969.

Rumer, Thomas A. *The American Legion: An Official History, 1919–1989*. New York: M. Evans, 1990.

Snell, Mark A. *Unknown Soldiers: the American Expeditionary Forces in Memory and Remembrance*. Kent: The Kent State University Press, 2008.

Sheffield, Gary. *Leadership in the Trenches: Officer–Man Relations, Morale and Discipline in the British Army in the Era of the First World War*. Basingstoke: Macmillan, 2000.

Sheffield, Gary. *Forgotten Victory: The First World War: Myths and Realities*. London: Review, 2002 [orig. 2001].

Thomson, Alistair. *ANZAC Memories: Living with the Legend*. Melbourne: Oxford University Press, 1994.

Todman, Daniel. *The Great War: Myth and Memory*. London: Bloomsbury, 2005.

Todman, Daniel. *Britain's War: Into Battle, 1937–1941*. London: Allen Lane, 2016.

Trout, Steven. *On the Battlefield of Memory: The First World War and American Remembrance, 1919–1941*. Tuscaloosa: The University of Alabama Press, 2012 [orig. 2010].

Winter, Jay. *The Great War and the British People*. Basingstoke: Macmillan, 1985.

Winter, Jay. *Sites of Memory, Sites of Mourning: The Great War in European Cultural History*. Cambridge: Cambridge University Press, 1995.

Winter, Jay. *Remembering War: The Great War between Memory and History in the Twentieth Century*. New Haven: Yale University Press, 2006.

Winter, Jay. "Demography." In *Companion to the First World War*, edited by John Horne, 248–262. Chichester: Wiley-Blackwell, 2010.

Wohl, Robert. *The Generation of 1914*. Cambridge: Harvard University Press, 1979.

Wootton, Graham. *The Official History of the British Legion*. London: MacDonald & Evans, 1956.

9 Between passatism and futurism

The rites of consecration to the Sacred Heart of Jesus in a transnational perspective (1914–1919)

Sante Lesti

Introduction: The past is a foreign country. They do things *slightly* differently there[1]

A historian of the First World War experiences a strange sense of familiarity reading the way in which François Hartog recently described the development, from the 1970s onwards, of a "new regime of historicity": that of "presentism".

> Has a somewhat different configuration [of time] not taken over since then, in which the distance between the space of experience and the horizon of expectation has been stretched to its limit, to breaking point? With the result that the production of historical time seems to be suspended. Perhaps this is what generates today's sense of a permanent, elusive, and almost immobile present, which nevertheless attempts to create its own historical time. It is as though there were nothing but the present, like an immense stretch of water restlessly rippling. So should we talk of an end of, or an exit from, modernity, from that particular temporal structure we call the modern regime of historicity? It is too early to tell. But we can certainly talk of a crisis. "Presentism" is the name I have given to this moment and to today's experience of time.[2]

With these very words – permanent, elusive, immobile – a historian of the First World War is accustomed to describing the present in which the societies involved in the Great War, and, in particular, the men in the trenches, were 'suspended'. This sensation of familiarity feels all the more heightened reading the words Paul Valéry used in 1935 to describe the way European societies experienced time in the aftermath of the First World War. Society's experience of time, as he explained, hung between two eras:

> On the one hand, there is the past that can neither be abolished nor forgotten, but from which we can derive almost nothing that will orient us in the present or help us to imagine the future. On the other hand, there is the future without the least shape.[3]

This notwithstanding, as John Horne explained in relation to the French soldiers (although this could be applied to all the main European armies), the future was not completely absent from the mental horizon of the combatants of the First World War. Certainly, from the autumn of 1914 to the summer of 1918 they experienced (and not only on the Western Front) an 'interminable present' in the form of trench warfare. But, they simultaneously maintained their hope in the 'dawn' of victory and peace.[4] From this perspective, the combatants of the First World War were suspended in a less oppressive present than our own, particularly in regard to what the weakest in our global society must endure: the precarious realities of workers, immigrants, exiles, and migrants.[5]

From whence does this (only partially) "false sense of familiarity with the past"[6] of the First World War come? We ourselves are experiencing, exactly in the same manner as the European societies involved in the First World War, a moment of "crisis of time" – that is, when "the way in which past, present, and future are articulated no longer seems self-evident".[7] Furthermore, and more specifically, today we are undergoing if not the same crisis, then *the same type of crisis:* the crisis of the modern regime of historicity. In Hartog's interpretation, this consists principally in the conception of history as process (and of time itself as the foremost actor in that process).[8] Unlike in the classical regime of historicity, summarised in the maxim '*historia magistra*', in the "modern regime of historicity" it is not up to the past to illuminate the present and the future, but up to the future – whether it's called Progress, or Nation, or Democracy, or Socialism, or European Union, to cite the main futurist ideology of post-1989 Europe – to make intelligible the past and the present, as moments in its history. Constructed progressively in the West – and through the influence of the West, in most of the world – during the course of the long eighteenth century, the "modern regime of historicity" experienced a definitive crisis, according to Hartog, during the 1970s and 1980s. It had passed through a former crisis, however, during and moreover after the First World War, as the writings of Paul Valéry, Stefan Zweig, Franz Rosenzweig, Walter Benjamin, and Gershom Scholem attest.[9]

The same category of 'regime of historicity' is a product of the temporal crisis we currently experience. It appeared – however "unobtrusive" this first appearance was[10] – in 1983,[11] the same year that *Time and Narrative, Volume 1* by Paul Ricoeur was published.[12] Hartog's research could mark its arrival in 2003, with the publication of *Regimes of Historicity: Presentism and Experiences of Time* [13] three years after, in this case, *Memory, History, Forgetting* by Paul Ricoeur.[14] In the meantime, during the 20 years between 1983 and 2003, time became a common theme both within academic literature and outside of it, in the fields of fiction and in public communications.[15]

The category of 'regime of historicity' is not simply a product of the crisis of time that we are passing through. It is one instrument for understanding the crisis of time that we are experiencing,[16] just as, more generally, the intervals in which the articulation of the past, the present, and the future "no longer seems self-evident".[17] The First World War undoubtedly constituted one of these intervals. This explains why the category of "regime of historicity" casts a new light on its cultural history.

Having thus clarified the origins of our methodology, just as, in general, the relationships between us, our words, and our object of investigation,[18] let us now concentrate on the subject under study: the rites of consecration to the Sacred Heart of Jesus celebrated in Europe during the First World War and its immediate aftermath (1914–1919). Why choose these particular rites? There are at least four reasons. The first is their centrality to the religious experience (and not only religious) of the societies involved in the conflict. These rites were of primary importance, not only for the many faithful who attended the ceremonies (tens of millions of Catholics in all Europe), but also for the way in which the rites were capable of raising expectations about the war. The second is the international spread of the rite, and in at least one case – the consecration of the Allied Nations (15 June 1917) – its transnational character. The third is that the "regimes of historicity" embedded in the rites of consecration permit a broader discussion of some aspects of the current understanding of the First World War as a moment of rupture in the individual and collective experience of time. The fourth is the coexistence of additional "regimes of historicity" at the centre of the same rites. In explaining this concept, Hartog wrote:

> Speaking of a "regime of historicity" is thus simply a way of linking together past, present, and future, or of mixing the three categories, in the same way that one talks of a "mixed constitution" in Greek political theory (combining elements of aristocracy, oligarchy, and democracy, one of which was always dominant in practice).[19]

As the rites of consecration to the Sacred Heart of Jesus celebrated in Europe during the First World War demonstrated, as every "regime of historicity" – whether presentist, passatist, or futurist – is the result of the combination of the three categories of past, present, and future, so every human experience of time is the result of the combination of two, or more, "regimes of historicity".

"The most important event of the whole war"

There were six main rites of consecration to the Sacred Heart of Jesus celebrated in Europe during the First World War and its immediate aftermath (1914–1919) among the belligerent nations: the consecration of Emperor Franz Joseph I of Austria and his lineage (8 December 1914); the national consecration of Germany (10 January 1915); the national consecration of France (11 June 1915); the consecration of the soldiers of the Italian army (5 January 1917); the consecration of the Allied Nations (15 June 1917); and the national consecration of France repeated on the occasion of the ceremony of the consecration of Sacré-Coeur (16 October 1919).[20]

The Archbishop of Vienna, Cardinal Friedrich Gustav Piffl, was the key instigator of the consecration of the Emperor and his lineage, the Habsburgs. The ceremony, officiated by the same Piffl, was celebrated on 8 December 1914, the day of the feast of the Immaculate Conception of the Blessed Virgin Mary, in the chapel of the Castle Schönbrunn, in the presence of all the members of the imperial court. On the first of January 1915, the consecration of the Emperor was repeated in all the churches of the Austro-Hungarian Empire.

As Claudia Schlager has argued, at the heart of the 'solemn' consecration of Germany were, instead, the Jesuit Father Karl Richstätter and Cardinal Felix von Hartmann, Archbishop of Cologne.[21] A *triduum* of penance preceded the ceremony celebrated on Sunday, 10 January 1915, in all the churches of the country. The national consecration of Germany was considered "the foremost religious event of the first years of the war" ("*das religiöse Zentralereignis der ersten Kriegsjahre*").[22]

On 11 January 1915, it was France's turn for consecration. In this case, Cardinal Léon-Adolphe Amette, Archbishop of Paris, organised the ceremony. The day before the consecration, celebrated in all the churches and chapels the nation over, the Archbishop of Besançon, Mons. Gauthey, proclaimed the ceremony "the most important event of the whole war" ("*la démarche la plus importante de toute la guerre*").[23] Of course, the Archbishop of Besançon was a long-time devotee of the cult of the Heart of Jesus. But like him, the whole episcopacy would in those weeks come to realise the exceptional importance of the event.

It was the Lord's own revealed commandment to a sister of the Convent of the Visitation of Salò (Brescia), Sister Maria Fedele Filippini (1859–1946), that spurred the Italian consecration of "all the army" to the Sacred Heart.[24] The consecration, organised by Father Agostino Gemelli and his entourage, was celebrated Friday, 5 January 1917, and involved two million soldiers, thanks as well to the double blessing of the military bishop, Mons. Angelo Bartolomasi, and Pope Benedict XV.

With the intention of extending the same "crusade for victory, peace, and Christian rebirth" to other nations of the Alliance, Father Gemelli called together, in April 1917, the Committee for the Consecration of the Allied Nations. The other national committees were presided over by various religious officials: for France, Cardinal Amette; for Belgium, the Rector of the Higher Institute of Philosophy at the Catholic University of Leuven, Mons. Deploige; and for England, by the Archbishop of Westminster, Cardinal Bourne. Two months later, on 15 June 1917, the consecration of the Allied Nations was carried out.

The cycle closed on 16 October 1919 with the consecration of France "penitent, devoted, and grateful" ("*poenitens et devota et grata*") celebrated at Montmartre (Paris) on the occasion of the dedication of the Basilica of the Sacré-Cœur. The national consecration was followed, on that occasion, by a solemn *triduum* comprising the feast of the Blessed Margherita Maria

Alacoque – whose visions were considered "the founding event in the epic tale of France and the Sacré-Coeur"[25] – on 17 October, a day of prayer for the war dead on the 18th; and finally a day of thanks for victory and peace on the 19th.

With the exception of the consecration of Francis I of Austria, we have here national consecrations, but not official ones – that is, those celebrated without the complicity of the political and military authorities of the states in question. Their goal, in fact, was to demonstrate to the political and military elites of these states (above all to the French Republic and the Italian Kingdom) the Catholic presence of these nations, thus opening the way for the re-Christianisation of the public sphere. By retracing the fundamental elements of the consecration ceremonies we can enter more directly into their contemporary experience of time.

A Christian regime of historicity?

During a second "exercise in viewing from afar" – that is, in viewing presentism from afar – Hartog takes into account "Augustine's phenomenology of human time", working from an outline of a Christian order of time. This latter is built, in Hartog's interpretation, on the "founding tension" between the decisive "already fulfilled" (the Incarnation) and the "not yet completed" (the Second Coming of Christ and the Last Judgment). The result is an "eschatological present", an "in-between time" of anticipation and vigilance.[26]

Do the rites of consecration to the Sacred Heart of Jesus reflect a Christian order of time? Notwithstanding the modern regime of historicity emerging during the long eighteenth century, they apparently do. In fact, with the exception of that of the soldiers of the Italian army, all the acts of consecration made reference to, in their opening, the decisive event of the Incarnation. Joseph I of Austria's consecration commenced with an invocation to "Lord Jesus Christ, son of the living God and king of heaven and earth".[27] The German Catholic pronouncement – none other than the *Act of consecration of the human race to the Sacred Heart* published by Leo XIII together with the encyclical *Annum sacrum*, in 1899[28] – opened with an invocation to "Most Sweet Jesus, Redeemer of the human race". The beginning invocation of the *Amende honorable et Consécration de la France au Sacré Cœur de Jésus* pronounced in all the churches and chapels of the country on 11 June 1915 referred not only to the decisive 'already fulfilled' (the Incarnation), but also to the 'not yet completed' (the Second Coming of Christ and the Last Judgment): "O Jesus, present and living in the Sanctified Sacrament of the Eucharist […]. We adore you our God and our Saviour, [you] who created us for good and for love to permit us one day to share in your eternal happiness."[29] A similar invocation was also made in the act of consecration of the Allied Nations (an unsurprising development, considering that its author was the same as that of the *Amende honorable et Consécration de la France au Sacré Cœur de Jésus*, that is, Cardinal Amette):

> Accept, O Lord Jesus, the offering that we make you [...]. We revere you as our God and We recognize you as our Lord, Sovereign and Master, because we were created from your Omnipotence and redeemed at the price of Your blood.[30]

The consecration of France (11 June 1915) and of the Allied Nations (15 June 1917), like that of the soldiers of the Italian army (5 January 1917), situated themselves within a dual reference to the 'already fulfilled' and 'not yet completed': on the one hand, the great revelation of 1689, in which the Lord had confided in Margherita Maria Alacoque his desire that France be consecrated to the Sacred Heart;[31] on the other, the fulfilment of the Lord's promise to give France victory over its enemies, and, more generally, the advent of the social kingdom of the Sacred Heart, or the fulfilment of a political order founded on the subordination of the state and society to God and his Church. Contrary to the Incarnation and the Last Judgment, the private revelations received by Margherita Maria Alacoque in the 1680s and the social kingdom of the Sacred Heart don't arrange the experience of time incorporated in the rites of consecration in any transcendental order. The result is a *secular* 'eschatological present', which superimposes – almost to the point of absorbing – the eschatological present of the Christian order of time. With this last development, the secular 'eschatological present' comprising the period between the great revelation of 1689 and the triumph of the social kingdom of Christ is an 'in-between time' of anticipation and vigilance, but also of an anticipation and a vigilance connected to the advent of a secular political order: "In anticipation of that happy day in which governments will bow before our God" says, for example, the incipit of the programme of the consecration of the Allied Nations published in the official *Bollettino* of the Italian Committee.[32]

As implied, in the case of the consecration of France, of the soldiers of the Italian army, and the Allied Nations, the Christian eschatological present tended to merge into the secular eschatological present generated by the Great Revelation of 1689 (the decisive already fulfilled) and the triumph of the social kingdom of Christ (the not yet completed). In the context of the Allied Nations' consecration, even the 1000 years' reign of Christ announced in the *Book of Revelation* finishes by coinciding with the social kingdom of Christ. At least, this is the case for a sermon by Don Angelo Roncalli, the future Pope John XXIII, who served at that time as a military chaplain.[33] Indeed, in the opening sermon of the novena of the feast of the Sacred Heart of Jesus, Don Roncalli cites the "magnificent" description of Jesus Christ riding a white horse contained in the *Book of Revelation* 19, 11–16[34] precisely in order to demonstrate the "foundations" of the "Divine Kingship of the Heart of Jesus".[35]

As Hartog has demonstrated, the expression 'order of time' dates back at least to the fifth century BC. Anaximander used it to suggest that things "suffer punishment and give satisfaction to one another for injustice according to the order of time".[36] For Herodotus, history was essentially (in Hartog's words) "the interval, calculated in generations, between an injustice and its punishment or redress".[37] However, such sequence of injustice and punishment characterises not only the Herodotean order of time, but also *another* Christian order of time – one that can be traced back to the Latin Fathers in their exegesis of Jesus weeping over Jerusalem. According to this particular Christian order of time, "the sins committed by reigns, cities, and families shall be punished in this life",[38] since there is no Last Judgment for them. On closer inspection, the rites of consecration to the Sacred Heart of Jesus seem more to reflect this other Christian order of time, based on the cyclic alternation of injustice and punishment, than that elaborated by Hartog. The only exception is the consecration of Franz Joseph I of Austria, which has no injustice to repair.

Let us begin with the *Amende honorable et Consécration de la France au Sacré Cœur de Jésus* pronounced in all the churches and chapels of the nation on 11 June 1915. As the title suggests, the first part of the act constituted an *amende honorable*; that is a public request of forgiveness for the sins of the French people and, moreover, of France as a nation – first and foremost the separation of the Church and the State.[39] According to Cardinal Hector-Irénée Sevin, Archbishop of Lyon, the consecration of France to the Sacred Heart not only remedied the post-revolutionary injustices committed by France – and in particular, the Third Republic – but restored to the nation its role as the "chosen people of the New Testament" – a role that it had claimed, despite the periods of 'guilt' that stained it, from the times of Saint Remigius.[40] Predictably, a solemn plea for the pardon of the committed sins also inaugurated the other formula of consecration drafted by Cardinal Amette, that of the Allied Nations. In fact, it says: "We implore of You to forgive the sins of those who have offered themselves to Your Heart", in the first part of the act.[41] Expiating the sins committed was the "first and foremost obligation" of the moment for the German bishops as well: "The war is punishment for all the people who are affected [...] Woe betide that people that not even this terrible scourge can induce to penitence: for they are destined to vanish." The sins of the German people were the same as the French, that is, the secularising laws brought to a culmination during the *Kulturkampf* first among all the institution of civil marriage.[42]

A passatist futurism[43]

At first glance, the rites of consecration to the Sacred Heart of Jesus celebrated in Europe during the First World War and its immediate aftermath (1914–1919) seem to be dominated by passatism; that is a regime of historicity dominated by the past.[44] As we have seen, from the past comes the Lord's injunction to France (but the injunction is interpreted by Italian Catholics as addressed to all people) to consecrate itself to his Sacred Heart, and from the past come the private and public sins to be absolved through a solemn request of forgiveness as well. Yet, it is another form of passatism that permeates the rites of consecration, namely the *historia magistra vitae.*[45]

Let us begin with the consecration of Franz Joseph I of Austria. In consecrating himself and his own dynasty to the Sacred Heart of Jesus, he believed that he was following in the footsteps of his predecessor Rudolf I of Germany (1218–1291), the Habsburg Emperor who first had showed, according to Franz Joseph, his love for the Sacred Heart.[46] The example mentioned by the German bishops in their invitational letter to all the Catholics of the *Reich* to participate in the national consecration was, if possible, even more authoritative. It referenced Constantine the Great: the example was taken directly from the encyclical of Pope Leo XIII *Annum Sacrum* (25 May 1899), in which, obviously, it was not the German Empire that was being compared to the Roman Empire, but Christianity as a whole:

> When the Church, in the days immediately succeeding her institution, was oppressed beneath the yoke of Caesars, a young Emperor saw in the heavens a cross, which became at once the happy omen and cause of the glorious victory that followed. And now, to-day, behold another blessed and heavenly token is offered to our sight – the most Sacred Heart of Jesus, with a cross rising from it and shining forth with dazzling splendour amidst flames of love.[47]

The very same example was cited by Cardinal Amette in his pastoral letter on the consecration of France (24 May 1915) – but with the following addition: "It was to the entire Church, to all people that this token of hope was revealed, this sign of victory and salvation. Yet, we could not forget that [...] it is above all to France that it was shown."[48]

Moreover, the private revelations bestowed upon Margherita Maria were not the only sign of Christ's privileged love for France. There were the victories of Clovis I, of Saint Louis, of Joan of Arc. From the past came, in other words, not only instructions as to how to confront the present (demonstrated from on high by God's divine mission for France, consecrated to his Heart), but also proof of the effectiveness of those instructions. This interpretation of God's will is evident in the words of Mons. Marbeau, the 'Bishop of the Marne':

> We have already received from those witnesses of his divine goodness that it is with the greatest trust that, the clergy and the faithful, we again address Him to beg for a swift [...] victory and peace [...] In these solemn festivities, We will also call upon all our Saints, protectors of France, because of their intercessional power with God, to secure for us the assistance that will permit our Nation to always stay faithful to the providential vocation that the soldier of God [is charged with]; the joyful fulfilment of God's plan through the Frankish *Gesta Dei per Francos.*[49]

From French history, ancient and recent, also came instructions for the behaviour of Catholic Italians:

> Devotion to the Heart of Jesus is the greatest hope of our time. We may obtain all, through the faith and love of the Heart of Jesus. He himself, appearing to the blessed Margherita Maria in France, said, "You will not lack my aid unless I lack power." Observe the French at the battle of the Marne: all appeared to be lost, when General Castelnau had the inspiration to invoke the Sacred Heart and consecrate the army. And the result was the marvellous victory that saved France.[50]

By the winter of 1916–1917, the miracle of the Marne was already a politically transnational legend.

The weight of the past notwithstanding, time, as understood by the men and women who attended the consecration ceremonies, was essentially (with the exception of the consecration of Franz Joseph I) a kind of futuristic experience. In effect, to give form to the present it was above all a vision of the future, if a nostalgic one: the social kingdom of Christ. Certainly, from the perspective of the authors of the consecration acts it was the past that modelled the present and the future. They believed not in the establishment, but in the re-establishment of the social kingdom of Christ: "We consecrate to you", one reads, for example in the *Amende honorable et Consécration de la France au Sacré Cœur de Jésus*,

> our people, all the force of our soul and our body: that they may be employed according to your will and belong to you completely! [...] To you we consecrate our families: we wish to give you reign through the observance of your commandments and of the teachings of your Church, so that fathers, mothers, and children truly belong to you [...] To you we consecrate, to the extent that we are able, our nation, promising you to work to *reestablish* (*rétablir*) your reign through the faith of your doctrine, submission to your law, and union with your Church: we desire that France belongs to you.[51]

Nevertheless, there was never any kind of social kingdom to restore, unless it was in a sort of mythical medieval past imagined from the vantage point of the future (the social kingdom of Christ to be established). In order to explain the oxymoron 'Conservative Revolution', Hartog wrote: "It is [...] none other than to mobilize the power [...] of the concept of revolution to freely recreate a past that never existed."[52] Despite all the differences between the two anti-liberal ideologies – the social kingdom of Christ and the "Conservative Revolution" – "To freely recreate a past that never existed" also encapsulates perfectly that which tens of millions of European Catholics involved in the rites of consecration to the Sacred Heart sought to achieve.[53]

Experiencing time from below

The analysis thus far has been mainly focused on acts of consecration, as well as – to a lesser extent – other sources from above, such as religious instructions, homilies, and chronicles. If we look at the same rites of consecration from below, we see a much more differentiated experience of time. Let us consider the case of the soldiers, about whom we know a little more. A minority of these seemed to entirely share the experience of time embedded in the acts of consecration. This is the case of the French soldiers gathered under the banner of *La Garde du Sacré-Cœur au front* ("The Guard of the Sacred Heart of Jesus at the Front"), an elite group of Catholic soldiers engaged in spreading the devotion to the Sacred Heart of Jesus among their fellow soldiers.[54] From their point of view, consecrating themselves and their nation (or their army) to the Sacred Heart of Jesus meant contributing to the advent of the social kingdom of Christ in the same way as it did for the authors of the different acts of consecration.[55] The members of *La Garde du Sacré-Cœur au front* fighting in the 3rd Alpine Battalion even called themselves "the Battalion of the Sacred Heart" ("*Le Battaillon du Sacré-Cœur*"):

> We, the Members of la Garde d'honneur de l'Insigne et du Drapeau du Sacré Cœur [*La Garde du Sacré-Cœur au front*], will do our utmost to make him [the Sacred Heart] beloved and respected here and therein our family and our dear France. He is a source of courage, strength, and virtue for us. Soon may we be the Battalion worthy of the Sacred Heart; the magnificent goal of your very beautiful oeuvre is to make him reign here and everywhere, and this will also be our goal, as our dear Chaplain often says.[56]

But these men were in the minority. For most of the soldiers, consecrating themselves and their nation (or their army) to the Sacred Heart of Jesus meant contributing not to the advent of the social kingdom of Christ, but to the advent of victory and peace, and/or the act of putting their body under the protection of the Sacred Heart of Jesus. As one Italian soldier put it, echoing the Constantinian motto '*In hoc signo vinces*': "With this sign [The Sacred Heart] I shall win."[57] Instead, one French soldier referenced to the scapular of the Sacred Heart of Jesus as "one of the most invincible shields" ("*un pare-balle, des plus invincibles*").[58] As stated, the two meanings often went together. For example, a French officer (probably a captain) deemed the banner of the Sacred Heart that his battery had just received from the chaplain to be a sign both of (bodily) 'protection' and victory.[59]

Each meaning resulted in a different experience of time. Or, if this is the case, in a combination of different experiences of time. For those who participated in the rites of consecration in order to put their body under the protection of the Sacred Heart of Jesus, there was no injustice to repair, or secular political order to be re-built. As a consequence, their experience of time does not reflect any Christian order of time, or any kind of futuristic experience. They were stuck in the present – not an eschatological present, but one made up of magical manipulations designed to help the soldier in question survive the trenches.

Even for those who participated in the rites of consecration with the aim of contributing to the advent of victory and peace, there was no injustice to repair. But there was an objective to be achieved, if not a secular political order to be re-built: victory and peace, as said. Despite this, their experience of time seemed to be more influenced by the past rather than the future, according to the classical regime of historicity summarised in the maxim '*historia magistra*'. By consecrating themselves and their nation (or their army) to the Sacred Heart of Jesus, they hastened, if not brought about, victory, exactly as Constantine the Great or Joan of Arc had done in the past.

Conclusion

According to a well-established tradition in First World War studies, the war constituted a major rupture in both the individual and the collective experience of time. This tradition dates back to Walter Benjamin, who, in a famous passage of his *The Storyteller* (1936), wrote:

> With the [First] World War a process began to become apparent which has not halted since then. Was it not noticeable at the end of the war that men returned from the battlefield grown silent – not richer, but poorer in communicable experience? What ten years later was poured out in the flood of war books was anything but experience that goes from mouth to mouth. And there was nothing remarkable about that. For never has experience been contradicted more thoroughly than strategic experience by tactical warfare, economic experience by inflation, bodily experience by mechanical warfare, moral experience by those in power. A generation that had gone to school on a horse-drawn streetcar now stood under the open sky in a countryside in which nothing remained unchanged but the clouds, and beneath those clouds, in a field of force of destructive torrents and explosions, was the tiny, fragile human body.[60]

There is no doubt that the First World War disaffirmed many kinds of military, economic, and cultural expectations, and broke life rhythms, especially the ones of rural communities.[61] Nonetheless, it would be seriously misleading to overlook the cultural practices by which individuals, communities, and institutions tried to re-establish continuity, and make sense of the war.[62] Among such cultural practices are the rites of consecration to the Sacred Heart of Jesus celebrated in Europe during the First World War and its immediate aftermath (1914–1919). Not only did they relate the present (the war) to the past (the providential history of the Sacred Heart of Jesus and the Habsburg dynasty, or France, or the human race) and the future (the advent of the social kingdom of Christ), but they also presented the war as a crucial turning point in this articulation between past, present, and future. At the same time, both the past and the future shed some light on the present, making it understandable – and possibly acceptable. In particular, the past acted in two ways: on the one hand, as a source of revelation – whichever the 'decisive already fulfilled' was, the Incarnation or the Great Revelation of 1689; on the other hand, as a collection of *exempla*, according to the model of '*historia magistra*'. In turn, the future (the advent of the social kingdom of Christ) provided a reference model.

The rites of consecration to the Sacred Heart of Jesus celebrated in Europe during the First World War and its immediate aftermath (1914–1919) allow us to point out new chronologies and timelines, both from an emic perspective (that is, the point of view of social actors) and from an etic perspective (that is, our own point of view).[63] From an emic perspective, the war was, as we have seen, a crucial turning point in the providential history of the Sacred Heart of Jesus and the Habsburg dynasty, or France, or the human race – a centuries-long history. From an etic perspective, the war has to be placed in the context of the Roman Catholic Church's reaction to the French Revolution and, in the shorter term, to the secularisation policies of the Third Republic, the Kingdom of Italy, and the German Empire.[64]

In his work "Entre expérience et mémoire: Les soldats français de la Grande Guerre", John Horne wrote:

> in their own way, the soldiers shared the illusions of the generals. They were themselves caught up in the mechanism of a future which seemed to implacably pass through the great liberating offensive. The hope that this time there would be an end to the war, that the German lines would finally crumble, renewed continuously, vacillating with the blackest despair in the face of an impossible breach.[65]

Among the cultural practices that continued to stoke illusions among the soldiers of an end to the war were the rites of consecration to the Sacred Heart: "When, moments later", one reads in the chronicle of the consecration of France celebrated in Sens on 11 June, 1915,

> the heads bowed under the blessing of the Sacred Sacrament [Holy Communion], it appeared that Our Lord extended reassurances that [...] he accepted the gift that France had just made him of herself, that he forgave her sins, and that he promised the nearness, with his powerful protection, of peace in a glorious victory. Thus comforted by this *hope* – or rather this *certainty* – the residents of Sens left that beautiful ceremony, and returned to their home to work towards the fulfilment of their promise: to make Jesus Christ the ruler of their hearts, their families, and their entire society, because France, re-emerging Christian, reoriented after decisive victory, would find itself first among nations.[66]

From the perspective of those who took part in the consecration rites, the ceremonies were an attempt to take hold of the future. From the perspective of the historian, the rites appear a most effective instrument for building up consensus.

Notes

1 This heading is of course an adaptation of the opening sentence of L.P. Hartley's novel *The Go-Between* (1953): "The past is a foreign country: they do things differently there."
2 Hartog, *Regimes of Historicity*, 17–18.
3 Valéry, *History and Politics*, 135–136.
4 Jagielski, "Modifications et altérations de la perception du temps"; Horne, "Entre expérience et mémoire". "The dawn" ["*L'Aube*"] is the title of the last chapter of Henri Barbusse's bestseller *Under Fire: The Story of a Squad (Le Feu)*.
5 "This presentist present is by no means uniform or clear-cut, and it is experienced very differently, depending on one's position in society. On the one hand there is the time of flows and acceleration, and of a valued and valorizing mobility, and on the other what the sociologist Robert Castel calls the 'status of casual workers [*le précariat*],' whose present is languishing before their very eyes, who have no past except in a complicated way (especially in the case of immigrants, exiles, and migrants), and no real future either (the temporality of plans and projects is denied them). Today's presentism can thus be experienced as emancipation or enclosure: ever greater speed and mobility or living from hand to mouth in a stagnating present. Not to forget a further aspect of our present: that the future is perceived as a threat not a promise. The future is a time of disasters, and ones we have, moreover, brought upon ourselves." Hartog, *Regimes of Historicity*, xviii.
6 "We constantly need to be shaken out of a false sense of familiarity with the past, to be administered doses of culture shock." Darnton, *The Great Cat Massacre*, 4.
7 Hartog, *Regimes of Historicity*, 16.
8 Hartog's account for the modern regime of historicity is based mainly upon Reinhart Koselleck. Koselleck, *Futures Past: On the Semantics of Historical Time.*
9 Hartog, *Regimes of Historicity*, 3–4.
10 Hartog, *Regimes of Historicity*, 8.
11 Hartog, "Marshall Sahlins et l'anthropologie de l'histoire".
12 Ricœur, *Time and Narrative, Volume 1.*
13 Hartog, *Regimes of Historicity.*
14 Ricœur, *Memory, History, Forgetting.*
15 Hartog, *Regimes of Historicity*, 2.
16 "My hypothesis (presentism) and my methodological instrument (the regime of historicity) belong together. The notion of a 'regime of historicity' helps shape the hypothesis of presentism, and the latter helps flesh out the notion of a 'regime of historicity'. The two are inseparable, at least in the first instance." Hartog, *Regimes of Historicity*, xv.
17 "I will use 'regime of historicity' sometimes in a broad, macrohistorical sense, sometimes in a narrow, microhistorical one. [...] Whatever the particular focus, I hope to generate new insights through close attention to moments of crisis of time and how these are expressed." Hartog, *Regimes of Historicity*, xvii.
18 Werner, and Zimmermann, "Penser l'histoire croisée: entre empirie et réflexivité".
19 Hartog, *Regimes of Historicity*, xv.
20 Lesti, *Riti di guerra.*
21 Schlager, *Kult und Krieg*, 139–156; Schlager, "Herz Jesu – ein Heldenkult?"

22 Busch, "Katholische Frömmigkeit und Moderne", 100.
23 *La Semaine religieuse du diocèse de Besançon*, 353.
24 Archivio Barelli, Dossier no. 26, Document no. 1, "Diario ricostruito attraverso le lettere di Padre Gemelli".
25 Jonas, *France and the Cult of the Sacred Heart*, 16.
26 Hartog, *Regimes of Historicity*, 55–63. Hartog's interpretation of the Christian order of time is built upon Cullman, *Salvation in History.*
27 *Sendbote des göttlichen Herzens Jesu*, 49–50.
28 Menozzi, *Sacro Cuore*, 212–224.
29 Archives Historiques de l'Archidiocèse de Paris, 3 G 2, 8, "Sacrè-Cœur = Consécration de la France".
30 Museo Storico Italiano della Guerra, Collezione Monterumisi, Santini, 0343, "Atto di consacrazione al Sacro Cuore di Gesù composto da S. E. il Card. Amette, Arcivescovo di Parigi per essere recitato la festa del Sacro Cuore 1917 dai popoli dell'Intesa per implorare la vittoria e la pace".
31 Benoist, *Le Sacré-Coeur de Montmartre: De 1870 à nos jours*, 187–198.
32 "La consacrazione delle nazioni alleate al S. Cuore", 38–39.
33 Butturini, "Servizio militare e impegni pastorali fra i giovani nell'episcopato di Marelli"; Roncalli, *Giovanni XXIII Angelo Giuseppe Roncalli*, 115–136.
34 "11. Then I saw the heavens opened, and there was a white horse; its rider was [called] 'Faithful and True'. He judges and wages war in righteousness. 12. His eyes were [like] a fiery flame, and on his head were many diadems. He had a name inscribed that no one knows except himself. 13. He wore a cloak that had been dipped in blood, and his name was called the Word of God. 14. The armies of heaven followed him, mounted on white horses and wearing clean white linen. 15. Out of his mouth came a sharp sword to strike the nations. He will rule them with an iron rod, and he himself will tread out in the wine press the wine of the fury and wrath of God the almighty. 16. He has a name written on his cloak and on his thigh, 'King of kings and Lord of lords'."
35 Archivio della Fondazione per le Scienze Religiose, Fondo Roncalli, CAM [Integrazioni dell'Archivio Roncalli] II, 299, "Novena del S. Cuore di Gesù in Duomo – giugno [1]917".
36 Hartog, *Regimes of Historicity*, 1.
37 Hartog, *Regimes of Historicity*, 1.
38 Celi, "Apologetica di guerra", 306.
39 "Nous vous demandons pardon des fautes, privées et publiques, par lesquelles nous avons outragé votre souveraineté et votre amour. [new line] Pardon, ô Seigneur Jésus, pour l'impiété qui voudrait effacer le nom de Dieu et votre nom béni de la face de la terre, et faire disparaître de partout votre croix, signe sacré de notre Rédemption. [new line] *Le Peuple*: Pardon, ô Seigneur Jésus! [new line] Pardon pour l'aveuglement et l'ingratitude de ceux qui, méconnaissant la mission divine confiée à votre Église pour le bonheur des sociétés non moins que pour le salut des âmes, ont voulu séparer d'Elle notre patrie et s'efforcent d'entraver sa liberté et son action parmi nous. [new line] *Le Peuple*: Pardon, ô Seigneur Jésus! [new line] Pardon pour la violation de vos commandements, pour les blasphèmes de parole et de plume, pour la profanation du Dimanche, pour le mépris des saintes lois du mariage, pour l'omission du grand devoir de l'éducation chrétienne, pour la dépravation des mœurs, pour l'amour effréné du luxe et du plaisir. [new line] *Le Peuple*: Pardon, ô Seigneur Jésus! [new line] Pour tous ces désordres, nous vous faisons amende honorable et nous vous demandons pardon." Archives Historiques de l'Archidiocèse de Paris, 3 G 2, 8, "Sacrè-Coeur, *Amende honorable et Consécration de la France au Sacré Cœur de Jésus*".
40 *La Semaine religieuse du diocèse de Lyon*, 19–20.

41 Museo Storico Italiano della Guerra, Collezione Monterumisi, *Santini*, 0343, "Atto di consacrazione al Sacro Cuore di Gesù composto da S. E. il Card. Amette".
42 *Lettera Pastorale dei Vescovi Tedeschi sulla Guerra*, 8–9.
43 It is worth noting that this definition has nothing to do with Italian Futurism. Instead, it has been elaborated starting from Hartog's categories of "passéisme" and "futurism". Hartog, *Croire en l'histoire.*
44 Hartog, *Croire en l'histoire*, 246.
45 Hartog, *Croire en l'histoire*; Hartog, *Regimes of Historicity*, 72–77.
46 *Sendbote des göttlichen Herzens Jesu*, 49–50.
47 *Lettera Pastorale dei Vescovi Tedeschi sulla Guerra*, 8–9; cf. Leo XIII (Pope), "Annum Sacrum", 25 May 1899, *The Holy See.* http://w2.vatican.va/content/leo-xiii/en/encyclicals/documents/hf_l-xiii_enc_25051899_annum-sacrum.html.
48 "Lettre de Son Eminence le Cardinal Archevêque de Paris au clergé et aux fidèles de son diocèse relativement à la consécration de la France au Sacré-Cœur", *La Semaine religieuse de Paris*, 786.
49 *La Semaine religieuse du diocèse de Meaux*, 291–292.
50 Archivio Segreto Vaticano, Segr. Stato, Guerra, fasc. 415, 2, f. 208r., "Solenne consacrazione dei soldati del R. Esercito Italiano al sacro Cuore di Gesù: Istruzione".
51 Archives Historiques de l'Archidiocèse de Paris, 3 G 2, 8, "Sacrè-Cœur, Amende honorable et Consécration de la France au Sacré Cœur de Jésus". "Nous vous consacrons nos personnes, toutes les puissances de notre âme et toutes les forces de notre corps: qu'elles soient employées selon votre volonté et soient ainsi totalement à vous! [new line] *Le Peuple*: A Vous, ô Seigneur Jésus! [new line] Nous vous consacrons nos familles: nous voulons vous y faire régner par l'observation de vos commandements et des préceptes de votre Église, afin que pères, mères et enfants soient vraiment à vous. [new line] *Le Peuple*: A Vous, ô Seigneur Jésus! [new line] Nous vous consacrons, autant que nous le pouvons, notre patrie, vous promettant de travailler à y rétablir votre règne par la foi en votre doctrine, par la soumission à vos lois, et par l'union avec votre Église: nous voulons que la France soit à vous. [new line] *Le Peuple*: Que la France soit à Vous, ô Seigneur Jésus!"
52 Hartog, *Croire en l'histoire*, 246.
53 It is worth noting that in the aftermath of the First World War the social kingdom of Christ (in the slightly different form of the Kingship of Christ) provided the theological basis for Pope Pius XI's attitude towards Italian Fascism: "Pius XI thus proposed a political theology constructed around a monolithic and authoritarian view of Catholicism. It was a system well-suited to an embrace with Fascism: for both shared an authoritarian, anti-liberal, anti-democratic and hierarchical root. But it was also a system that, precisely by the virtue of its all-inclusive character, could not but vie with the totalitarian claims of the State that Mussolini was about to construct" (Ceci, *The Vatican and Mussolini's Italy*, 77–78). See also Menozzi, *Sacro Cuore*; Menozzi, "Spiritualità del regno e politica nel secondo dopoguerra"; Menozzi, "La dottrina del regno sociale di Cristo tra autoritarismo e totalitarismo"; Menozzi, "Regalità di Cristo e politica nell'età di Pio XI"; Menozzi, "Ernesto Balducci e il regno sociale di Cristo".
54 "La Garde du Sacré-Cœur au front", 2.
55 Archives Historiques de l'Archidiocèse de Paris, Basilique du Sacré Cœur de Montmartre, 4 C 23, "Culte du S[acré] C[œur] au front. Guerre de 1914–1918–1939–1940. Lettres d'Aumôniers et de Soldats – (et 1956)".
56 Archives Historiques de l'Archidiocèse de Paris, Basilique du Sacré Cœur de Montmartre, 4 C 23, "Culte du S[acré] C[œur] au front", 3. "Membres de la Garde d'honneur de l'Insigne et du Drapeau du Sacré-Cœur nous ferons tout notre possible pour qu'il soit aimé, honoré ici d'abord, dans notre famille plus tard et dans

notre chère France. Oui il est pour nous source de courage, de force et de vertus. Puissions-nous être bien vite le digne bataillon du Sacré-Cœur le faire régner ici et partout c'est le but magnifique de votre œuvre si belle ce sera la nôtre comme nous le répète souvent notre cher Aumônier."

57 Stiaccini, "Con questo segno vinco", 956. "L'unione fa la forza, e ha chi / sta ben unito col nostro buon Dio / non perisce in eterno / e con questo segno + vinco / Viva Gesù, Viva Maria, / Evviva S. Damiano. / Addio."

58 Archives Historiques de l'Archidiocèse de Paris, Basilique du Sacré Cœur de Montmartre, 4 C 23, "Culte du S[acré] C[œur] au front", 3.

59 Archives Historiques de l'Archidiocèse de Paris, Basilique du Sacré Cœur de Montmartre, 4 C 23, "Culte du S[acré] C[œur] au front", 7.

60 Jay, "Walter Benjamin, Remembrance and the First World War", 232.

61 While there is doubt that the war marked a transition from a 'traditional' to a 'modern' memory, as asserted by, among others, Fussell, *The Great War and Modern Memory*, and Jedlowski, "Memoria, mutamento sociale, modernità".

62 Mazzini, "Cose de laltro mondo".

63 Ginzburg, "Our Words, and Theirs: A Reflection on the Historian's Craft, Today", *Cromohs* 18 (2013), www.fupress.net/index.php/cromohs/article/view/14122/13124.

64 Clark and Kaiser (eds), *Culture Wars: Secular–Catholic Conflict in Nineteenth-Century Europe.*

65 Horne, "Entre expérience et mémoire", 903–919.

66 *La Semaine religieuse de Sens et Auxerre*, 371–373.

Bibliography

Unpublished material

Archives Historiques de l'Archidiocèse de Paris, 3 G 2, 8, "Sacrè-Cœur".

Archives Historiques de l'Archidiocèse de Paris, Basilique du Sacré Cœur de Montmartre, 4 C 23, "Culte du S[acré] C[œur] au front. Guerre de 1914–1918–1939–1940. Lettres d'Aumôniers et de Soldats – (et 1956)."

Archivio Barelli (Milan), Dossier no. 26, Document no. 1, "Diario ricostruito attraverso le lettere di Padre Gemelli".

Archivio della Fondazione per le Scienze Religiose (Bologna), fondo Roncalli, CAM [Integrazioni dell'Archivio Roncalli] II, 299, "Novena del S. Cuore di Gesù in Duomo – giugno [1]917".

Archivio Segreto Vaticano, Segr. Stato, Guerra, fasc. 415, 2, f. 208r., "Solenne consacrazione dei soldati del R. EsercitoItaliano al sacro Cuore di Gesù: Istruzione".

Museo Storico Italiano della Guerra (Rovereto, Trento), Collezione Monterumisi, Santini.

Published material

Barbusse, Henri. *Under Fire: The Story of a Squad (Le Feu)* [*Le Feu: Journal d'une escouade*]. Translated by Fitzwater Wray. New York: E.P. Dutton & Co., 1917.

Benoist, Jacques. *Le Sacré-Cœur de Montmartre: De 1870 à nos jours*, vol. 1. Paris: Les Éditions Ouvrières, 1992.

Busch, Norbert. *Katholische Frömmigkeit und Moderne: Die Sozial- und Mentalitätsgeschichte des Herz-Jesu-Kultes in Deutschland zwischen Kulturkampf und Erstem Weltkrieg*. Gütersloh: Gütersloher Verlagshaus, 1997.

Butturini, Lucia. "Servizio militare e impegni pastorali fra i giovani nell'episcopato di Marelli". In Angelo Giuseppe Roncalli (Giovanni XXIII). *Nelle mani di Dio a servizio dell'uomo: I diari di don Roncalli 1905–1925*. Bologna: FSCIRE, 2008.

Ceci, Lucia. *The Vatican and Mussolini's Italy*. Translated by Peter Spring. Leiden: Brill, [2013] 2017.

Celi, Gervasio. "'Apologetica di guerra': Giustizia e misericordia." *La Civiltà cattolica*, 67, no. 3(1916): 186–197 and 298–311 (article is in two parts).

Clark, Christopher and Wolfram Kaiser (eds). *Culture Wars. Secular–Catholic Conflict in Nineteenth-Century Europe*. Cambridge: Cambridge University Press, 2009.

Cullman, Oscar. *Salvation in History*. Translated by Sidney G. Sowers. New York: Harper & Row, 1967.

Darnton, Robert. *The Great Cat Massacre: And Other Episodes in French Cultural History*. New York: Basic Books, 1984.

Fussell, Paul. *The Great War and Modern Memory*. Oxford: Oxford University Press, 1975.

Ginzburg, Carlo. "Our Words, and Theirs: A Reflection on the Historian's Craft, Today." *Cromohs* 18 (2013). www.fupress.net/index.php/cromohs/article/view/14122/13124.

Hartog, François. "Marshall Sahlins et l'anthropologie de l'histoire." *Annales: Économies, Sociétés, Civilisations* 38 (1983): 1256–1263.

Hartog, François. *Croire en l'histoire*. Paris: Flammarion, 2013.

Hartog, François. *Regimes of Historicity: Presentism and Experiences of Time*. Translated by Saskia Brown. New York: Columbia University Press, [2003] 2015.

Horne, John. "Entre expérience et mémoire: Les soldats français de la Grande Guerre." *Annales: Histoire, Sciences Sociales* 60 (2005): 903–919.

Koselleck, Reinhart. *Futures Past: On the Semantics of Historical Time*. Translated by Keith Tribe. New York: Columbia University Press, [1979] 2004.

Jagielski, Jean-François. "Modifications et altérations de la perception du temps chez les combattants de la Grande Guerre." In *La Grande Guerre: Pratiques et expériences*, edited by Rémy Cazals, Emmanuelle Picard, and Denis Rolland, 205–214. Toulouse: Privat, 2005.

Jay, Martin. "Walter Benjamin, Remembrance and the First World War." In *Walter Benjamin: Critical Evaluations in Cultural Theory*, II, *Modernity*, edited by Peter Osborne, 230–249. London: Routledge, 2005.

Jedlowski, Paolo. "Memoria, mutamento sociale, modernità." In *La memoria contesa: Studi sulla comunicazione sociale del passato*, edited by Anna Lisa Tota, 40–67. Milan: Franco Angeli, 2001.

Jonas, Raymond. *France and the Cult of the Sacred Heart: An Epic Tale for Modern Times*. Berkeley: University of California Press, 2000.

"La consacrazione delle nazioni alleate al S. Cuore." *Bollettino mensile dell'Opera di consacrazione nazionale al sacro Cuore di Gesù* 1 (30 April 1917).

"La Garde du Sacré-Cœur au front." *Le Pèlerin* 42 (11 August 1918).

La Semaine religieuse de Sens et Auxerre 52 (19 June 1915).

La Semaine religieuse du diocèse de Besançon 40 (10 June 1915).

La Semaine religieuse du diocèse de Lyon 22 (4 June 1915).

La Semaine religieuse du diocèse de Meaux 48 (29 May 1915).

Leo XIII (Pope). "Annum Sacrum." 25 May 1899. The Holy See. http://w2.vatican.va/content/leo-xiii/en/encyclicals/documents/hf_l-xiii_enc_25051899_annum-sacrum.html.

"Lettre de Son Eminence le Cardinal Archevêque de Paris au clergé et aux fidèles de son diocèse relativement à la consécration de la France au Sacré-Cœur." *La Semaine religieuse de Paris* 62 (29 May 1915): 786–790.

Lesti, Sante. *Riti di guerra: Religione e politica nell'Europa della Grande Guerra*. Bologna: il Mulino, 2015.

Lettera Pastorale dei Vescovi Tedeschi sulla Guerra. Firenze: Libreria Editrice Fiorentina, [1914] 1915.

Mazzini, Federico. *"Cose de laltro mondo": Una cultura di guerra attraverso la scrittura popolare trentina 1914–1918*. Pisa: Edizioni ETS, 2013.

Menozzi, Daniele. *Sacro Cuore: Un culto tra devozione interiore e restaurazione cristiana della società*. Rome: Viella, 2001.

Menozzi, Daniele. "Spiritualità del regno e politica nel secondo dopoguerra." In *Parola in dialogo: Scritti in onore di Paolo Ricca e Sergio Rostagno*, 283–298. Torino: Claudiana, 2003.

Menozzi, Daniele. "La dottrina del regno sociale di Cristo tra autoritarismo e totalitarismo." In *Cattolicesimo e totalitarismo: Chiese e culture religiose tra le due guerre mondiali (Italia, Spagna, Francia)*, edited by Daniele Menozzi and Renato Moro, 17–55. Brescia, Morcelliana, 2004.

Menozzi, Daniele. "Ernesto Balducci e il regno sociale di Cristo: tradizione e rinnovamento nell'ideologia politico-religiosa del cattolicesimo novecentesco." In *Storia ed esperienza religiosa. Studi per Rocco Cerrato*, edited by Alfonso Botti, 139–160. Urbino: Quattroventi, 2005.

Menozzi, Daniele. "Regalità di Cristo e politica nell'età di Pio XI: i congressi internazionali di Cristo Re." In *Chiesa, laicità e vita civile. Studi in onore di Guido Verucci*, edited by Lucia Ceci and Laura Demofonti, 153–172. Roma: Carocci, 2005.

Ricœur, Paul. *Time and Narrative, Volume 1*. Translated by Kathleen McLaughlin and David Pellauer. Chicago: University of Chicago Press, [1983] 1984.

Ricœur, Paul. *Memory, History, Forgetting*. Translated by Kathleen Blamey and David Pellauer. Chicago: University of Chicago Press, [2000] 2004.

Roncalli, Marco. *Giovanni XXIII Angelo Giuseppe Roncalli: Una vita nella storia*. Milano: Mondadori, [2006] 2007.

Schlager, Claudia. *Kult und Krieg: Herz Jesu – Sacré Cœur – Christus Rex im deutsch-französischen Vergleich 1914–1925*. Tübingen: Tübinger Vereinigung für Volkskunde e. V. Schloß, 2011.

Schlager, Claudia. "Herz Jesu – ein Heldenkult? Emotionsgeschichtliche Perspektiven auf Maskulinisierungsstrategien einer populären katholischen Frömmigkeitsform im Umfeld des Ersten Weltkrieges." *Schweizerische Zeitschrift für Religions- und Kulturgeschichte/Revue suisse d'histoire religieuse et culturelle* 108 (2014): 241–257.

Sendbote des göttlichen Herzens Jesu 51 (February 1915).

Stiaccini, Carlo. "Con questo segno vinco: La religiosità popolare nelle testimonianze dei soldati della Grande Guerra. " In *Humanitas*, 48, no. 6(2008): 943–958.

Valéry, Paul. *History and Politics*. Translated by Denise Folliot and Jackson Mathews. New York: Bollingen Foundation, 1962.

Werner, Michael, and Bénédicte Zimmermann. "Penser l'histoire croisée: entre empirie et réflexivité." In *De la comparaison à l'histoire croisée*, edited by Michael Werner and Bénédicte Zimmermann, 15–49. Paris: Seuil, 2004.

10 Hoping for victorious peace

Morale and the future on the Western Front, 1914–1918

Alex Mayhew[1]

> Wars themselves [...] are conducted with the intention of peace, even when they are conducted by those who are concerned to exercise their martial prowess in command and battle. Hence it is clear that peace is the desired end of war. For every man seeks peace, even in making war; but no one seeks war by making peace.
>
> (Saint Augustine[2])

Conflict, peace, and the passage of time were interwoven in the minds of soldiers during 1914–1918. Men, whose worlds became dominated by sedentary war, sought to project forward momentum onto the conflict. This chapter stems from a wider research project that focuses on the morale of English infantrymen serving on the Western Front. It will argue that men's morale was, in part, sustained by their hope for peace and that they constantly imagined, visualised, planned, or fantasied about this future they constructed internally. This counters existing literature, which tends to view thoughts of peace as an indicator of low morale, disaffection, and war weariness. Ultimately, it demonstrates how victorious peace became the focus of men's desires and positively influenced their morale. They came to perceive and sense agency as a member – no matter how insignificant – of an organisation they saw as the vehicle through which peace would be achieved. It also suggests that soldiers did not see victory in strategic terms, but perceived it as the key to a brighter future. Their frames of reference – the boundaries that constrained and focused their perceptions of the world – meant that the war's positive progress became the key to their peaceful future.

A well-known satirical advert in *The Somme Times* poked fun at men's widespread investment in this seemingly distant future, and apparently mocked their optimistic outlooks. It asked the following questions of its readers:

1 Do you suffer from cheerfulness?
2 Do you wake up in a morning feeling that all is going well for the Allies?

3 Do you sometimes think that the war will end within the next twelve months?
4 Do you believe good news in preference to bad?
5 Do you consider our leaders are competent to conduct the war to a successful issue?

If musing upon this, the reader found that he had answered "yes" to one or more of these questions, the advert consoled him that "we can cure you of this dread disease".[3] All it took, it seems, was a visit to the front line. Some commentators have used this extract as evidence that the "sojourn of the trenches" had degraded the British soldiers' positivity to the point where optimism was "a severe affliction" and enthusiasm "a deadly sin".[4] However, this is to fundamentally misunderstand the nature of soldiers' self-reflection. Optimism was an attitude incumbent on the men fighting in Belgium and France – it was necessary for their emotional survival.[5] Indeed, only five months later the same publication's editorial played on a similar theme, this time berating pessimism. The pessimist, it claimed, was a "strange elfish creature".[6] Perhaps in a similar vein, Private Albert Joy kept a signature book, in which family, friends, and comrades would leave messages. One note, from J. Baker, another soldier in the Royal Fusiliers, advised Bert to "despond if you must. But never despair for remember that from our greatest failures arise our most Brilliant success."[7] Men were aware of optimism's irony, particularly in the face of the Great War's bitter and prolonged struggle, but they remained resolutely forward looking and their perceptions of the future were integral to this and to their morale.

Morale has been the focus of many studies of the First World War. Scholars have sought, in particular, to understand how men endured the horror, the exhaustion, and the boredom of 1914–1918.[8] A number of 'key terms' revolve around the topic – for instance, consent, coercion, and, more recently, compliance – but a working definition of morale is highly contested. These range from S.L.A. Marshall's encapsulation of morale as the entire 'thinking' of an army,[9] to scholars who maintain a more functional focus on 'carrots' and 'sticks'.[10] Nevertheless, more recent studies have had greater success in developing a functional definition that also embraces the complexity and multifaceted nature of this phenomenon. Alexander Watson sees morale "as the readiness of a soldier or a group of soldiers to carry out the commands issued by military leadership",[11] while Jonathan Fennell defines it as "the willingness of an individual or group to prepare for and to engage in an action required by an authority or institution".[12] These two definitions, though, suggest that soldiers were only reacting to the military's demands. Morale also relies on a constructive mind-set, in which positive visions of a productive future are integral. Morale, in this instance, is defined as the process by which men, positively or negatively, rationalised their role as soldiers and constructive members of the military.

Peace is approached here from the perspective of the individual soldier – as a subjective (and constructed) future space, in which the war has ended, and their worldly desires have been realised. Visions of such peace proved to be a sustaining factor in morale. However, there were limits to this. In August 1917, III Army Censor, Captain M. Hardie, reported that the soldiers' letters he had analysed "give little indication of a wish for 'peace at any price,' but they do show an immense and widespread longing for any reasonable and honourable settlement that will bring the war to a close".[13] He went on to suggest, "there is a feeling of uncertainty as to the progress of our arms to an ultimate victory".[14] There is much debate as to the state of British morale during this period, some historians arguing that the months leading up to the German offensives of 21 March 1918 were characterised by an ebbing endurance among British soldiers.[15] Peace had been a prevalent theme in previous reports, yet the experience of 1917 had left men desperate for its arrival. In previous years, Hardie had been more confident about the troops' resilience despite their deep-seated desire for the conflict's cessation. A year earlier he had recorded: "the necessity of passing a third winter in the trenches is cheerfully faced, and there is a general acceptance of the idea that the War will not be over for another year".[16] Nonetheless, the British Army recovered from this pessimism and the despondency engendered by the experience of Passchendaele did not fatally undermine the army. It was, in fact, rarely seen at other times during the war.[17] Indeed, the term '*après la guerre*' was used regularly by the men in Belgium and France, where it became "a magical phrase used by soldiers [...] longing for survival and for the return of peace".[18]

The relationship between soldiers' hope for peace and morale is a relatively unexplored facet of their experiences in the Great War. The importance of peace to them has been underlined by a number of scholars, who mainly focus on how it became a negative ingredient in soldiers' morale. Paul Fussell, for one, argued that by late 1916 "the likelihood that peace would ever come was often in serious doubt".[19] Helen McCartney has also demonstrated how the shocking experience of battle could induce a temporary desire for "an almost unconditional peace".[20] Elsewhere the thirst for peace has been analysed in relation to war weariness. Jean-Jacques Becker discussed how 1917 saw an increase in French citizens' "desire for peace". In particular, significant minorities began to support a compromise peace and some even called for peace "at any price".[21] Peace, then, could be an unreachable goal and was often a product of people's desire to escape war. Such conclusions were undoubtedly true in the context of, for example, Russian soldiers in 1917. However, John Horne has revealed how French soldiers' concepts and hopes for peace provided visions of a future and past that allowed men to cope with the reality of their present.[22] What follows will extend such an analysis and argue that historians should focus on the constructive and beneficial dimensions of soldiers' preoccupation with their peaceful futures.

Men's "first concern was to survive".[23] As such, they hoped and yearned desperately for war's end. The word 'hope' emerges from the letters and diaries of the men who lived and died on the Western Front and is pervasive in secondary accounts.[24] Some argue that the "emphasis of morale" enables an individual "to live and work hopefully and effectively".[25] Studies of morale in war tend to reference hope without assessing its impact or adequately defining it. Where the term is used, it is often in passing. For example, Jonathan Boff has discussed German soldiers' hope for improved rations and an early peace in late 1918 and Alexander Watson underlines soldiers' hope for survival, nurtured by periods of rest.[26] Vanda Wilcox has looked at soldiers' passivity, citing Agostino Gemelli, an Italian military psychologist, who argued in 1917 that successful soldiers actually benefited from being hopeless as it allowed them to adapt to war.[27] Ben Shephard has suggested that British soldiers on 1 July 1916 were "buoyed up with hope and excitement, the men went calmly and uncomplainingly to their deaths".[28] Yet its lack of definition makes it difficult to understand the nuances and functions of hope. Some social psychologists assert that hope is an abstract emotional process at the extreme end of positivity. Barbara Fredrickson, for example, sees hope as an antidote to fear. She claims that:

> hope is not your typical form of positivity. Most positive emotions arise when we feel safe and satiated. Hope is the exception. It comes to play when our circumstances are dire [...] Hope literally opens us up. It removes the blinders of fear and despair.[29]

Such an encapsulation of hope might ally itself with Watson's hypotheses regarding "positive illusions" and "optimistic reasoning" being parts of the soldiers' psychological coping mechanisms.[30] Often inflated and unrealistic appreciations of chances of survival, for example, facilitated endurance. Equally, intense hopefulness for peace perhaps motivated men to suffer war without morale collapsing.

However, many other psychologists argue that 'hope' is primarily a cognitive process and not an emotion and is indeed different from optimism and positivity, which can affect people's hope.[31] They contend that hope is goal orientated. Optimism and positivity influence the way in which one rationalises a situation and relate to expectations of the future.[32] An optimistic explanatory style facilitates "higher levels of motivation, achievement, and physical well-being and lower levels of depressive symptoms".[33] In such studies "hope is defined as goal-directed thinking in which people perceive that they can produce routes to desired goals (pathways thinking) and the requisite motivation to use those routes (agency thinking)".[34] Barriers can block the path to their attainment:

> When encountering barriers that impede goal pursuit, people appraise such circumstances as stressful. According to the postulates of hope

> theory, positive emotions result because of the perceptions of successful goal pursuit. Conversely, negative emotions typically reflect the perceived lack of succcss [...] Thus, their perceptions regarding the success of goal pursuits casually drive subsequent positive and negative emotions... Furthermore, these emotions serve as reinforcing feedback.[35]

In this instance, positivity and optimism – and their antitheses – operate within the wider boundaries of hope. Thus, with this definition, hope is a cognitive goal-driven process that can be affected by emotional reactions – such as anger, despondency, disappointment, or confidence – to obstacles that arise in the pursuit of these objectives.

This analysis looks to throw light on soldiers' hopes and visions of peace. It begins by discussing how visions of peace emerged on the Western Front, before discussing exactly what 'hope' means in a psychological context. It argues that soldiers' hope became focused on victorious peace, which formed around visions of a future devoid of war. It will highlight the role of hope in English soldiers' morale – underlining the ways in which internalised visions of the future were prevalent and contrasted with and combated the men's present. It will examine men's desire for peace across the war and consider how this interrelated with morale, before arguing that their frame of reference – the war and the Western Front – constrained their perspective on the world. This ensured that the war's military progression and peace became intimately related in the minds of men.

Memories, fantasies, and visions of peace

How did these visions of peace emerge and what form did they take? The war very quickly became a route march. Men's pasts lay at their point of departure, their present was immersed in the war, but their futures – the destination – would see peace return to their lives.[36] This became the ultimate hope of men who fought in France and Belgium in 1914–1918 and seems to have developed remarkably quickly. Herbert Trevor, a subaltern at the onset of war, was left overwhelmed and disenchanted by the retreat from Mons and the Battle of Le Cateau. In a postcard on 22 September 1914 he expressed solidarity with the despondent German prisoners he had met, remarking that "everyone would like peace".[37] Men throughout the conflict shared this sentiment.[38] Yet peace was an abstraction and a construct – fed by memories, dreams, and fantasies of home. It formed in the minds of individual men and, as such, it would be impossible to provide any valid 'average' vision of peace. Age, regional background, class, and any number of other factors influenced what men saw when they imagined peace. It is, however, possible to trace similarities among the soldier's visions of the future. The soldiers' own words show that it is possible to find continuity in attitude across the conflict. It is apparent, from a close reading of men's interpretation of their situation, that both officers and men were conscious of the horrors and lack of romance in modern war. Yet they retained a belief that it was something to be borne.[39] This general desire for peace helps to explain the apparent solidarity of soldiers in the pursuit of, as some vocalised, a common object. These men believed that this objective stemmed from shared experience.[40] Given the encapsulation of 'hope' as goal orientated it is necessary to understand what this aspiration might have been.

These ideas and visions of peace were internalised and allowed men to escape their present. Many soldiers' journals provide the historian with insights into how and when these images would appear. Importantly these tended to focus on the home or close community and not only provided sustenance but were also a reminder of whom these men were fighting for. These 'dreams of home' emerged from sleep and daydreams and in any quiet moment the men might have had for reflection. Many contemporary postcards aimed at soldiers played on these themes. A great number depicted men (either in cartoon or staged photograph) sitting, sleeping, or writing in recreated trench scenes – a smaller image would depict their vision of home. Wives, sweethearts, and mothers were among the more popular objects of longing.[41] The age demographics of the army, which saw many of the men in the New Armies being under the average age of marriage, means that Michael Roper is probably correct in his assertion that the maternal link was often the strongest.[42] Yet, it was by no means the only bond with the United Kingdom. The rank and file tended to focus on their family or friends. L/Cpl. Cecil White recorded in his diary how he had "the pleasure" of receiving "a letter from Tom" and noted that his "thoughts are continually with them [his family] and all my desires for peace so that we may all return to our homes". White logged how he and his comrades continued to "spend our days with those we love".[43] R.E.P. Stevens' visions took him from the dining-room fireside, via "the parish of Godly [and] every street" to the "interiors of houses I was want to visit all at once".[44] Officers' social networks, which were often broader than those of the ranks, seemed to have produced a more diverse image of peace. The *Cambridgeshire Territorial Gazette* described how a soldier slumped "mud-caked, hungry and tired" into an "inviting armchair" behind the lines. Removing his boots and tunic, he "gazed into the glowing fire and was on the point of conjuring up visions of 'leave' and London".[45] Authors in *The B.E.F. Times* explored similar themes. One poem, entitled "Brazier Pictures", began "In my brazier as I gaze / Pictures come and pictures go, / Dimly seen across the haze – / Christmases we used to know". The soldier in question's imagination was "burning clear", allowing "visions of better days" and eradicating "discomforts that are near" and as "pictures come and pictures go" he was fortified.[46] Other contributors described how dawn's light brought visions of their "dearest" in her "silver shrouded dress" or how "the evening mists" became a "portrait".[47] Officers were also occupied by images of "green fields," "peace," and drinking in their clubs.[48]

These images emerged in the form of inner visions of domestic peace and they provided an escape. They all focused on peaceful scenes, in which the war had been removed. The static and disempowering nature of warfare of the Western Front meant that men were – in a quasi-meditative way – dwelling on the past, and fantasising about the future, so that they could ignore their presents.[49] Men were not exclusively obsessed with their pasts *or* their presents, but imagined both as they sought to combat the stresses of their day-to-day lives at war. Many of these visions of past and future were intentionally domestic and focused on facets of the peaceful world that had been removed from the Western Front. Women, an unscarred countryside, or a bustling townscape were individualised but played a similar role from soldier to soldier. Cecil White highlighted the collective benefits of these visions: "for a time the gloom and squalor of the filthy trench disappeared".[50] It was in such moments that time became fused: memories of the past fed visions of the future, which – even for a moment – allowed men to escape the conflict's horrors. Importantly, even when men focused on memories rather than fantasy, this was used to construct a perfect post-war world. This was integral to morale.

Hope, peace and morale

For this idealised future to ever materialise, it was necessary for soldiers to successfully endure and traverse their presents. As Alexander Watson has argued, "human faith, hope and optimism, no less than cultural traits, discipline, primary groups and patriotism, explain why and how men were willing and able to fight in the horrendous conditions of the Western Front".[51] The importance of hopes for future peace in reference to morale rested on its ability to mitigate the trauma of war. Its primary function in this respect is its capacity to expand men's mental horizons. Hope is imbued with a sense of forward momentum. A man, sitting in a muddy, water-filled hole in the ground, with shells exploding in the near vicinity and the recent vision of a friend maimed or killed imprinted on his memory, needed to invest himself in thoughts of a brighter future.[52] For some men, these visions provided some warmth, as they pondered England, "home" and "beauty".[53] Any inability to conceive of the future, in moments of extreme tension, played negatively upon the minds of some soldiers.[54] These visions also benefited those living through the mundane routines of life behind the line. Peace provided an alternative world. Hope and hopes of peace allowed men to visualise themselves outside of their present situations, at least until they were forced to "wake up".[55] The soldiers' imaginings of their lives after war and encapsulations of a better time to come are important in understanding how and why men were able to endure the horrors of 1914–1918.

Peace simultaneously provided a universal and unifying goal for the men, and a set of aspirations that could become highly personal. Reverend E.N. Mellish told his old parish that the men he encountered "longed for home".[56] More explicitly, L/Cpl. C. White wrote in his diary that "peace is the only event that is worth carrying on for".[57] Peace plans offered men the opportunity to develop personal life goals. These usually involved the desire to be reunited with loved ones – children, spouses, parents, and friends.[58] Or they might involve education, their occupation, embracing an England not devastated by war, simply sitting at home, sleep, good meals, pastimes, their 'land fit for heroes' or grandiose ideas regarding the betterment of society. Peace was a concept and construct that meant different things to different people; as such it meant that they were likely to have been devoted to their imagined future.[59] Importantly, the ability to rationalise peace on such personal levels may very well have instilled a depth of meaning that was sustaining. As Captain Hardman related to his parents, his vision of the "dawn of peace" contrasted with the horror, despondency, and suffering. It reminded him that "I have something to live for beyond this world of carnage [and] bloodshed".[60] As far as research can demonstrate, humans are the only meaning-seeking species. Thus, it makes sense for men to have personalised war and allied themselves with a hope for life after it.[61] Lieutenant Brettell and his sweetheart looked to peace as an opportunity to profit from his war experiences. They discussed selling his paintings, which depicted the difficulties of life in the army, and making "pots of money".[62] The process of having a goal in which meaning is constructed allows people to "construe their lives as meaningful or worthwhile".[63] Simply, a hope for, and thoughts of, peace could imbue men's war experience in general with a sense of purpose that went some way to diminishing its horrors and facilitating allegiance to the military and victory.

Peace offered the opportunity for men to plan the consolidation of all that was good in their lives and encouraged them to live with a vision of this possibly idealised future. Importantly, the use of the future tense diminished their present and, by envisaging their role in these scenarios, assumed their survival. The end of the war offered the opportunity for happiness. 2nd Lieutenant Henrick Jones, 1/7 Royal Warwickshire Regiment, exemplified the benefits of romantic relationships. He was engaged in December 1916 and the belief in his and his fiancée's promised future together had not diminished by March 1918. He exclaimed, "when all this is over we are going to have the most wonderful time".[64] A certainty permeated his statement. This might have represented an attempt to allay his fiancée's fears. Yet, the fact that he was forced to talk of their future in such a tone cannot have failed to have had an impact on him. This was not limited to men's love lives. The collection of correspondence kept by A & C Black Publishers provides another example of the ways in which men focused on peace. Their employees wrote with surprising regularity, discussing their jobs, thanking their employers for their provisions to the families left at home, enquiring about the business and discussing their employment upon their return to civilian life.[65] Future plans might have been as simple or mundane as rebuilding or designing one's house or, most pervasively, even just being at 'home'. In S.B. Smith's correspondence with his wife he spent much of his time consoling her that, while he was physically absent, his thoughts and feelings were very much focussed on "home". He was insistent that he was pining for a return to this place of security and happiness.[66]

The hope for peace allowed men to envisage their future selves outside and beyond the confines of their dangerous military service. There was a self-confidence that, while perhaps unjustified, played an important psychological tool. The ability to imagine future episodes is an important facet of human cognition.[67] Psychologists Cristina Atance and Daniela O'Neill describe this process as "episodic future thinking" and have pointed towards its potential in positively affecting human experience and behaviour.[68] Importantly, research reveals that *writing* about these goals can lead to significant improvements in subjective well-being. Writing about "one's best possible self" has similar constructive benefits to writing about trauma and can increase general wellness. The benefits of these processes are more powerfully felt when these future goals are used in contrast to shocking events.[69] This may have also provided men with a coping strategy that, through fantasy and imagining the future, allowed the soldiers to deflect fatalism and have become a ritual through which they avoided considering their own fragile mortality. The very process of hoping, thinking, and placing themselves in a future situation, and the vocalisation of these visions, provided men with a psychological tool by which they were able to go some way in protecting themselves from the horror – and indeed boredom – that stalked much of their present.

However, for most of the war any peace was still very distant – and false hope might have engendered disillusionment.[70] Every year saw men pondering the proximity of peace. In 1914, while some historians of the home front have argued that ideas of the war ending before Christmas quickly disappeared, soldiers in France and Belgium certainly retained these thoughts.[71] The sentiment of L.F. Ashburner in October 1914, is something that recurred across the period and the different British armies. He asked the question, "I wonder if we shall be home by Xmas?"[72] Similarly, in late October 1916, Private William Anderson told his wife in a letter that "I haven't lost the idea that this business may well be over by Christmas. Maybe my faith or hope is on the strong side."[73] Again, at the beginning of 1918 many were certain that the war's intensity and destruction could not last for more than another year. Christmas and New Year provided focal points for soldiers' aspirations for peace. This might be a product, in the case of Christmas, of religiosity. Interestingly, however, there is an absence of references to Easter in this regard, perhaps because the festive holidays were more focused on family and friends. December became a marker for the coming of peace. This suggests that men wished to create a sense of certainty that the war would end and reflects a general desire to avoid winter in the trenches.[74] More fundamentally, it is evidence that many men viewed peace as the product of military campaigning. Their hope was that a victory would have been gained by Christmas, a festival that comes in the month after campaigning usually ended. G.H. Greenwell, for one, typified this. He argued that "everything sooner or later has an end" and hoped that a successful campaign would see them out of the trenches by Christmas.[75] In an earlier letter he attempted to reason that a German collapse would be engineered by that year's campaign. He pointed (prematurely) to the entry of "Roumania [sic]" as a decisive blow to the Central Powers, the result was that he "shouldn't be surprised at a sudden German collapse".[76] R.D. Sheffield also reported that, in November 1914, the men around him "all seem to say that they think this will be the last battle and if we win that the war will end before Christmas".[77] In a world dominated by conflict, men were hopeful that the current campaign would be their last. Christmas and New Year became the symbolic waypoint for this success. However, such hopes, once dashed, risked leading to disenchantment and undermining the benefits of being invested in peace.

Yet, what of hopes that were more easily undermined and discredited? Commentators have discussed the negative consequences of false illusions, inappropriate goals, and poor strategies for the attainment of these goals. Could it be the case that being hopeful might have worked to the detriment of motivation and therefore morale? Research suggests that even false hope can still be of psychological benefit.[78] If hope is "a positive motivational state that is based on an interactively derived *sense of success* (a) agency (goal-directed energy), and (b) pathways (planning to meet goals)" this might make sense.[79] "Hope [...] is influenced by a dispositional sense of abilities to produce pathways and agency across situations."[80] Thus, motivation might be influenced to a greater extent by the perceived success (or failure) in the spheres of agency and pathway, rather than attainment. It is not necessarily the goal that sustains motivation but the feeling of being on a legitimate course to the said goal. Indeed, it would seem that high-hopers have a slight, but not counterproductive, tendency to view their reality with a "positive self-referential bias".[81] Some of the more insightful notes that refer to victory are revealing in their description of men who were "under the impression that they are winning". Men were willing to ally themselves with *feelings* of success.[82] It is also possible that the nature of a close knit unit that intentionally fosters *esprit de corps* might have facilitated these high levels of hope, research suggesting that interpersonal relationships are of huge benefit to the processes of resilient hoping.[83] These relationships certainly existed; Charles Carrington, for one, saw the friendships that he had developed – with both officers and men – as one of the key aspects of his war experience. He even feared the severance of the bonds that he felt the war had created.[84] It is likely that resilience was a by-product of such solidarity. Importantly, it was the hopes for things such as better weather – over which men, clearly, had no control – that left them most despondent. During the winter it was the weather, not the war, that became most unbearable and, in fact, men looked to battle as an opportunity to end the war before the onset of winter weather. The war was controllable; the patterns of nature were not.[85] Therefore, so long as men were told or felt that the war was progressing positively, it would be through this that they measured success.

Their hopes for peace relied on a sense of agency – and morale benefited from this. The pursuit of a goal and the cognitive function of hoping for peace provided many men with at least the perception of forward momentum. It would seem that as soldiers they considered it their duty to follow orders and that in doing so felt that they were playing some small part in ensuring that a victorious peace might emerge that bit faster. As one man recalled, he and his comrades were motivated by a belief that "so long as we are soldiers and orders are orders and must be obeyed".[86] Another man made such an assertion earlier in 1915: "an Army Order is an Order, and has to be obeyed".[87] The sense that a good soldier should be obedient is explicit here, but implicit is a belief. that this was also their best course of action. However, it might be, as John Brophy and Eric Partridge recalled, that soldiers quickly "grew used" to army rules and "obeyed without thinking".[88] Indeed, 'O'Grady', a game similar to 'Simon Says', was popular among soldiers at rest precisely because it played on the fact they were "conditioned to an instantaneous and mindless response to drill orders bawled out with authority, to stop and think first was not so easy".[89] There were, however, orders that were considered unnecessary and were ignored – particularly when it came to unnecessarily disturb the peace on a quiet front in midwinter, or when personal comfort was at stake.[90] Interestingly, Alexander Watson has found that, after the German Offensives in 1918, British soldiers told their German captors they were more confident in *their own* ability than the generals'.[91] It might have been that by fulfilling their role as a soldier to the best of their ability men were, in part, developing both agency and self-confidence. For the most part, obedience was considered a necessary duty – by conforming to orders soldiers were, somewhat perversely, taking an active role in the successful orchestration of the war.

There were more overt examples of men seeking and celebrating agency. Ghassan Hage, an anthropologist studying migration, has said: "it is when people feel that they are existentially 'going too slowly' or 'going nowhere', that is, that they are somehow 'stuck' on the 'highway of life', that they begin contemplating the necessity of physically 'going somewhere'".[92] By assuming the role of the soldier and contributing to the smooth running of the military machine, soldiers were playing a small part in the war's successful orchestration. In believing that they were treading a route towards their ultimate goal of peace – however slow that path might have been – soldiers allied themselves to constructive activity. This is evident in men's descriptions and celebration of visible or reported German prisoners. It was hard to find agency or evidence of progress in the midst of static warfare. This made casualties and prisoners part of an attritional tit-for-tat. W.J. Lidsey proudly reported on 3 November 1916 that the battalion his unit was relieving had captured 700 enemy combatants but had only suffered "very slight" casualties.[93] Similarly Sergeant William Summers related, in a triumphant tone, that he and his "chaps" had captured an old German machine gun as a "war trophy".[94] It is likely that this explains one of Haig's "lessons" of the Somme Campaign; the "morale effect" of aerial superiority had been "out of all proportion" to its material impact.[95] Men were receptive to and regularly emphasised any signs of success. After 21 March 1918 open warfare was celebrated – even in retreat – as a sign that the war offered new opportunities.[96] This supports psychologists' arguments that, in many instances, "the process of *doing* […] is itself the goal".[97] Soldiers desperately wanted evidence, any evidence, of the war progressing positively. So long as there was some way to paint the war encouragingly it seems that men were willing to do so. It is here where the importance of human endurance, underlined by Alexander Watson, is so important.[98] Watson has highlighted this as an integral component of the story of morale in the First World War. The nature of the human psyche meant that men were able to continue fighting and believing in its ultimate efficacy despite setbacks. In the atmosphere of static war, it is certainly true that feeling that you were 'going somewhere' was often highly beneficial. Yet, how, in the context of sedentary war, were men able to relate peace to the military reality?

Military victory as the 'pathway' to peace

It is necessary to investigate what seems to have contributed to soldiers' continuing vision of victory as the route to peace. Good morale – in the functional sense – relied on men's motivation to fight and to follow orders. In a world in which their frames of reference was constrained by context – that of Belgium, France, and the BEF – men were offered little opportunity to view their situation objectively. Ultimately, many men sensed that military victory could provide the instrument for peace and that the armed forces were the vehicle through which this goal could be achieved.

A number of factors influenced men's perception of potential avenues to peace. The relationship between hope and religion should not be underestimated.[99] Believing that forces outside one's comprehension control the world might have led men to toe the line.[100] Religion and hope are intimately connected and recent historiography has looked to overturn previous arguments regarding the increasing secularism of Britain in the years before 1914. British society at this time, if not overtly religious, maintained a high level of internal religiosity.[101] Even during periods of the most intense warfare, where possible Sunday church service remained a notable part of a battalion's weekly schedule.[102] Furthermore, research suggests that the presence of clergy in the front lines and aid stations was widespread.[103] This fed a heavy use of religious imagery, closely related to ideas of heaven, in soldiers' poetry and writings. Reverend E.N. Mellish's visions of peace saw a time when "the rain is over and gone, the flowers appear on the earth, [and] the time of singing birds is come".[104] Peace was Eden. In contrast the Western Front came to represent a man-made "hell on earth".[105] The physical manifestation of this can be seen in pieces of trench art. There are examples of soldiers' craft that use metal, sometimes bullets, shell casings, or even the remnants of zeppelins to fashion crosses or objects imbued with religious symbolism.[106] The

Christian symbolism usurped the destructiveness of these objects. Entrusting events to God influenced men's vision of the world and of their war experience – if only in that it facilitated an appreciation of a brighter future, entrusted to a higher power. For example, A.J. Lord, in a letter to his parents, mused about waiting for "the Dove of Peace".[107] Similarly, William Anderson looked to the same Christian symbol, asking for it to "descend and make the World a brighter and happier place".[108] Their faith seems to have engendered thoughts of peace and they entrusted its deliverance to God.[109] Scholars have pointed towards the potent relationship between Christian 'utopia' and 'hope' – "hope driving the utopian impulse, utopianism inspiring hope".[110] Postcards, too, were produced with religious motifs that were meant to assuage fears for the future. E. Grantham kept a postcard entitled 'faith' in his wallet for the duration of the war. This card exhorted its reader to have faith that a "brighter" future would come.[111] It might also be that prayer, as an act, nurtured some agency as men beseeched God to interfere with the progression of the war on their behalf.[112] All of this contributed to a willingness to place trust in God, fuelling some men's belief that the military path, while horrific, was to be endured as a part of something greater than themselves. There was a certainty that this would ultimately draw them closer to a utopian 'tomorrow'.

There is also the sense in which the soldiers' perception of the Germans undermined any belief in peace being attainable by a path other than victory. While hate and desire for revenge fuelled reactions to the death of comrades, this generally occurred momentarily during the aftermath of battles or bombardments. English soldiers' perception of German expansionism, injustice, and exploitation of the civil populace did much more to encourage ideas of a fought-for peace.[113] It could remind men of the vulnerability of the home they were fighting for.[114] In 1914 the soldiers were conscious of Belgium's plight and had recent memories of both civilian casualties and of the retreat from Mons.[115] By 1916, whilst on the offensive, they were aware of the German operations at Verdun.[116] Finally, in 1918, men seemed to begin fearfully anticipating a German onslaught as early as February. 21 March confirmed their fears for home as they again witnessed a German advance, its dangers and the resultant dislocation of the local populace.[117] Importantly, the men could picture the Germans as "the invader".[118] The significance of this characterisation cannot be understated; it meant that war did not seem futile. German strategy reminded men that they were fighting a destructive enemy who was seemingly incapable of compromise. Many, like S.A. Newman, collected and retained sets of postcards, widely available in France (in both English and French), which showed the destruction wrought upon towns such as Arras. In these images the "Hun's passion for destruction" was emphasised and the "devastation by the vampires" powerfully portrayed.[119] The men regurgitated this characterisation of their enemy in their own writing: Sgt. S. Gill noted, in scrawled handwriting, "damn those Prussians. Sodom and Jeremiah to the [...] demonical ghoulish vampires."[120] It was possible to see the

Germans as the "authors" of the evil wrought on the world and this, in many cases, engendered an "intense determination to exact full retribution".[121] What is more, it made any peace without Germany's defeat seem futile. Of course, by the Hundred Days Campaign a confidence in the likelihood of retribution was, in itself, sustaining.

The men's characterisation of the German state and military hierarchy compounded this. Outside the chaos of battle there is very little to suggest that the majority of men maintained an obvious hatred for the enemy soldiers as individuals.[122] However, there is evidence that, in the eyes of the soldiers, Germany's leaders were villains. They were seen as an evil presence towards both the civilised world and their own soldiers and citizens. This was vividly portrayed in Bairnsfather's illustration "The Tactless Teuton". In this drawing a stick-thin German private of the "Orphans' Battalion" was being overseen by an overweight and moustachioed man of his own Army's "Gravediggers' Corps" – a caricature typical of many others. Bairnsfather presented a state-run slaughter orchestrated by a devilish "Hun".[123] This "powerful ideological" motivation, which had characterised the German state as an evil aggressor, was repeated by soldiers in their depictions of the enemy.[124] A.J. Lord described to his parents how "with that wonderful thoughtfulness for which he is noted, the Hun saved up his offensive for my first morning in the Line on returning from leave!"[125] The use of 'Hun' or 'Teuton' in the letters and diaries of front-line soldiers both dehumanised the enemy and suggested a timeless, negative, and expansionist militarism. If the ordinary German soldiers could not normally incite hatred, this abstract idea certainly could. This is reflected in a man's short letter to his employer in England: "[I] heard all about the raid, it's a bad business and shows what a callous lot the Huns are […] They are only creating rage in the hearts of the chaps out here."[126] More explicitly, Lt. Reginald Neville explained to his sister that:

> We want to exterminate, not so much the Germans as individuals for they are harmless enough, but their methods and principles; and worst of it is that there is only one way to do this and that is to kill the individuals. If only we could get hold of these abstract principles, turn them into concrete and then blow them off the face of the earth, our end would be attained.[127]

Germany became a historicised warmonger and it made sense that this enemy should be met with military might. You cannot parley with a remorseless and bellicose entity. Furthermore, it is easier to kill a monster than a man. The nature of the German state in the minds of British soldiers facilitated a belief that peace could only be the product of victory.

Close analysis of the soldiers' letters reveals the ways in which their frames of reference influenced their perception of the route to peace. Both their present, and their physical and social environments, constrained their ability to rationalise the world. They were members of the military and this identity was driven to the fore, and while they imagined a future outside of the military, the army was the only way in which they could make sense of this future. Hew Strachan sees this negation of the self as a key ingredient in training's influence upon morale.[128] There is a sense – within letters in particular – that a number of 'voices' operated in two distinct contexts. Men, who were immersed in war, wrote to their loved ones, who were sitting at home. There is an illustration of the men's present and future here. Men were transported, mentally, in their moments of both reading and writing letters, to the locality of their writer and recipient. This might have entailed as simple a process as imagining the weather at home.[129] In many ways it was this connection that facilitated their ability to envisage and become so attached to peace. Yet, they simultaneously asserted their soldierly identity. Foucault's discussion of authorship is useful in understanding this; he argued that "the author function operates so as to effect the dispersion of these [...] simultaneous selves".[130] These men, while pining for peace and embracing their future identity in a peaceful world, revealed their militarisation in their use of language – they immersed their syntax in indirect and often unintentional indications that they were at war, which represented the 'present' tense for them. Their conversations with home justified fighting; they asserted that they are doing their "duty as a soldier".[131] Less vocal but equally revealing examples are provided by the regularly used slang. Words such as 'straffe' or 'straffing', 'wind-up', 'A1', 'grousing', and 'napoo' quickly entered the vocabulary of every soldier. Importantly these were used to describe most situations.[132] This seems to have influenced the way that their loved ones interacted with them. Military language permeated letters sent to soldiers from the home front, where the language – even of wives and sweethearts – adopted a militaristic tone. Lieutenant Brettell's love interest, Peggy, said that she would like to "straffe" a number of people in England.[133] The war and the military embedded itself in the psyche of some men and their loved ones and would have certainly played a part in their conceptualisation of peace.

Soldiers became immersed in their environment and this influenced or even dominated their perception of the world. A non-military route to peace was fundamentally inconceivable. Mark Hewitson has observed that "although soldiers often felt themselves to be an increasingly important part of a national 'history', they were also increasingly alienated from [...] external and official history, turning in on themselves".[134] This is reflected in Charles Quinnell's revelation that he and his comrades came to see themselves as a "race apart from the civilians".[135] This was most evident to Quinell when he was home on leave, in the moments where he had been removed from the military environment. There is a sense in the recollections and writing of Quinnell and others that the army had become, if only temporarily, a key aspect of their existence. Correspondingly Neitzel and Welzer have argued that conceptions of duty and war among men of the Wehrmacht between 1939 and 1945 were "firmly anchored in [a] soldiers' frame of reference".[136] The language of a number of the English soldiers between 1914 and 1918 suggests that many men were unable to seriously contemplate peace other than through battle and victory. As soldiers the army was the vehicle for action, however small their role in that process. As a result some envisaged, somewhat apocalyptically, peace as the product of a 'great battle', and the vast majority saw peace as the essential result of military triumph – defeat was not mentioned as a possible route to this ideal.[137] For many, victorious peace was the only possible, or conceivable, outcome. Whether this reflects the success of the state and army's attempts to link peace and victory in the minds of its soldiers, or more simply stemmed from a basic psychological defence mechanism (or, indeed, was a combination of the two), requires further consideration and research.

Conclusion

Visions of peace existed throughout the war, a conflict that saw men invest themselves in an ideal future. Their desire for peace was an understandable reaction to the novelty of death on such a scale – a basic comprehension of war resulting in death could not have prepared them for the overwhelming and pervasive nature of what Audoin-Rouzeau and Becker have termed "total battle".[138] An example of this is C.S. Baines' narrative of 31 October 1914, the opening day of the First Battle of Ypres. He describes a conversation with an officer of the battalion to his left, who related that his unit had been "overwhelmed".[139] The use of this word is interesting; Baines' Battalion War Diary reports that there had been withdrawals but makes no reference to any unit being overcome by the enemy. It is possible, then, to view "overwhelmed" as indicative of the officer's and his men's emotional reaction to battle. Indeed, he told Baines that if his men had not been wounded by the shellfire, then they had been "shattered" psychologically.[140] Four years later, in the days before the Armistice, men still found it necessary to comment on the terrible extremities of war, one diary describing it simply as "massed slaughter and destruction".[141] The psychological impact was felt after combat and influenced the focus of men's goal orientation. However, hopes and visions of a peaceful future provided a mechanism to divert or diminish the negative impact of war's trauma by expanding their mental horizons and imbuing events with some redeeming qualities. Humans craft narratives to makes sense of the world, and peace (in particular) allowed men to visualise a personalised brighter future. This, in some way, justified their sacrifices and the images these thoughts conjured contrasted with the war and assured their survival. The very process of hoping and investing oneself in the future was of psychological benefit and undoubtedly benefited morale in this regard.

Yet, myriad issues influenced soldiers' views of the route – or pathway – to this ultimate goal. To understand the durability of men's belief in military victory's ability to bring peace, it is necessary to look at the realities of war and the language of the men. This insists that while peace was their supreme wish, the nature of life at the front engendered a general inability, once immersed in the military, to see beyond the warpath. While war was frequently horrendous and hard to bear, their soldierly outlook dominated many men's impressions of the world and their conception of Germany's despotism meant that peace seemed unlikely – or less desirable – without the defeat of Kaiser Wilhelm's regime. Men wanted peace and home, but were in a world immersed with the imagery and reality of war. As such, their appreciation of their existence and the route to peace was often constrained by their military boundaries. This is likely to have facilitated a willingness to follow orders – adherence to military authority was often the only conceivable or certain ingredient for peace.[142] It pushed their combatant identity to the fore. This and soldiers' very negative perceptions of the German state as a militant aggressor were potent ingredients in dictating how men interpreted the war and their role in it. This lies at the heart of how and why men retained a belief in military victory, which came to represent their most likely route to the peaceful future they desperately hoped for.

Notes

1 I would like to thank Dr Heather Jones, Professor David Stevenson, Professor Sönke Neitzel, Dr Joanna Lewis, and the anonymous reviewers for their comments on previous drafts of this chapter.
2 Brown, Nardin, and Rengger (eds), *International Relations in Political Thought*, 126.
3 National Army Museum (NAM) 1959-03-34 (4) "Are You a Victim of Optimism?", *The Somme Times: with which are incorporated The Wipers Times, The "New Church" Times & The Kemmel Times*, No. 1 Vol. 1 (Monday 31 July 1916).
4 Tholas-Disset and Ritzenhoff, *Humor, Entertainment, and Popular Culture during World War I*, 2.
5 Watson, *Enduring the Great War*, 92–107.
6 NAM 1959-03-34-1 "Editorial", *The BEF Times*, No. 2 Vol. 1 (Monday 25 December 1916).
7 NAM 1992-09-139-6 Papers of A. Joy, Signature and Message by J. Baker in Signature Book.
8 Watson, *Enduring the Great War*, 2008; Watson, "Morale"; Sheffield, *Command and Morale*; Kitchen, *The British Imperial Army in the Middle East*; Bowman, *The Irish Regiments in the Great War.*
9 Marshall, *Men Against Fire*, 158.
10 Ferguson, *The Pity of War.*
11 Watson, "Morale", 176.
12 Fennell, *Combat and Morale in the North African Campaign*, 9; Wilcox, *Morale and the Italian Army during the First World War*, 4.
13 Imperial War Museum (IWM) 84/46/1 Private Papers of M. Hardie, Report on III Army Morale 1917.
14 IWM 84/46/1 Private Papers of M. Hardie, Report on III Army Morale 1917.
15 Beckett, Bowman, and Connelly, *The British Army and the First World War*, 153–154; Stevenson, *With Our Backs to the Wall*, 267–268; Watson, *Enduring the Great War*, 184; Boff, *Winning and Losing on the Western Front*, 92–122; Englander, "Discipline and Morale in the British Army, 1917–1918", 141.
16 IWM 84/46/1 M. Hardie, "3rd Section Report on Complaints, Moral, Etc. (1916)".
17 IWM 84/46/1 M. Hardie, "Report on III Army Morale January 1917".

18 Brophy and Partridge, *The Long Trail: Soldiers' Songs and Slang 1914–18*, 66.
19 Fussell, *The Great War and Modern Memory*, 71.
20 McCartney, *Citizen Soldiers: The Liverpool Territorials in the First World War*, 214.
21 Becker, *The Great War and the French People*, 217–235.
22 Horne, "Entre expérience et mémoire: Les soldats français de la Grande Guerre", 903–919.
23 Fuller, *Troop Morale and Popular Culture in the British and Dominion Armies*, 63.
24 IWM P 229 Private Papers of Brig Gen. H.E. Trevor, Postcard to Evelyn Parker 22 September 1914; IWM 96/24/1 Private Papers of Pte. W.M. Anderson, Letters to Wife 9, 17, 18, and October 1916; IWM 98/28/1 Private Papers of Pte. F.G. Senyard, Letters to Wife 5, 17, 26 December 1916, 18 September, 28 June and 10 July 1918; Royal Fusiliers Museum Archive (RFM) RFM.ARC.3032 L/Cpl. C. White, Diary 7, 10 January, 8, 27 February, 5, 7, and 8 March 1917.
25 Brotemarkle, "Development of Military Morale in a Democracy", 79.
26 Boff, "The Morale Maze: The German Army in Late 1918", 11; Watson, "Morale", 187.
27 Wilcox, "Weeping Tears of Blood", 174.
28 Shephard, *A War of Nerves: Soldiers and Psychiatrists 1914–1994*, 41.
29 Frederickson, "Why Choose Hope?", *Psychology Today*, 23 March 2009. www.psychologytoday.com/blog/positivity/200903/why-choose-hope.
30 Watson, *Enduring the Great War*, 146.
31 Lopez, Snyder, and Pedrotti, "Hope: Many Definitions, Many Measures", 91–107.
32 Carver and Scheier, "Optimism", 75.
33 Reivich and Gillham, "Learned Optimism: The Measurement of Explanatory Style", 57.
34 Reivich and Gillham, "Learned Optimism: The Measurement of Explanatory Style", 94.
35 Reivich and Gillham, "Learned Optimism: The Measurement of Explanatory Style", 94.
36 IWM 84/46/1 M. Hardie, "3rd Section Report on Complaints, Moral, Etc. (1916)".
37 IWM P 229 Brig General H.E. Trevor, Postcard 22 September 1914.
38 NAM 2005-02-6 Papers of Capt. Maurice Asprey, Letter Mother 27 September 1914; IWM 06/5/2 Private Papers of Brig. General G.A. Stevens, Letter to Mother 11 November 1916; IWM 77/33/1 Private Papers of Lt. J.H. Johnson, Diary 28 and 31 December 1916; NAM 7403-29-486-144 Papers of Sgt. Harry Hopwood, Letter 12 October 1917; IWM 01/21/1 Papers of H.T. Madders, Diary 14 October 1918.
39 IWM 06/5/2 Brig. General G.A. Stevens, Letters October through December 1916.
40 IWM 99/13/1 Private Papers of Major S.O.B. Richardson, Letter [No. 4] Late 1916.
41 Soldiers of Oxfordshire Museum (SOFO) Box 16 Item 61 Collection of 15 Patriotic CARDS 1914–1918 Vintage, Postcard: "La Reve"; Liddle/WW1/GS/0746 Papers of H.O. Hendry, Postcard: "I'm Thinking of You [No. 504]"; Manchester Regiment Archive MR 2/17/65 Papers of John Douglas Powell, Postcard: "Dreams of Home [No. 8]"; MR 3/17/126 Papers of Pte. John Peat, Postcard: "Sketches of Tommy's Life. Out on Rest – No. 9".
42 Roper, *The Secret Battle: Emotional Survival in the Great War*, 59.
43 RFM.ARC.3032 L/Cpl. C. White, Diary 12 February 1917.
44 IWM 02/43/1 R.E.P Stevens, Diary 6 December 1916.
45 British Library – *Cambridgeshire Territorial Gazette*, No. 4 (1916), 88.
46 British Library – "Brazier Pictures", *The B.E.F. Times*, No. 2, Vol. 2 (1916).
47 British Library – "The Sybarite's Soliloquy", *The B.E.F. Times*, No. 2, Vol. 2 (1916), 8; "To My Marraine", *The BEF Times*, No. 3, Vol. 2 (1917).
48 "Disturbing Influences", *The BEF Times*, No. 3, Vol. 2 (1917), 3.
49 M. Roper, "Nostalgia as an Emotional Experience in the Great War", The Historical Journal 54, no. 2 (June, 2011), 440.

50 RFM.ARC.3032 L/Cpl. C. White, Diary 19 February 1917.
51 Watson, *Enduring the Great War*, 107.
52 IWM 77/33/1 Lt. J.H. Johnson, Diary 18 August 1916.
53 RFM.ARC 2013.8.2 Pte. P.N. Wright, [Retrospective Diary] 10 November 1916.
54 SOFO Box 16 Item 29 3/4/J3/8 Lt. J.E.H. Neville, Diary 23 March 1918.
55 Liddle/WW1/GS/0583 Papers of S. Frankenburg, Letter 1 December 1917.
56 IWM PP/MCR/269 Private Papers of Reverend E.N. Mellish, Letter in St Paul's, Deptford, Parish Church Magazine (July 1918).
57 RFM.ARC.3032 L/Cpl. C. White, Diary 16 January 1917.
58 IWM 08/66/1 Private Papers of 2nd Lt. D. Henrick Jones, Letters to Wife 14 October 1916 and 4 April 1918; IWM 98/28/1 Pte. F.G. Senyard, Letter to Wife 28 August 1918.
59 Fletcher, *Life, Death and Growing Up on the Western Front*, 3.
60 IWM 76/27/3 Private Papers of Col. Hardman, Letter to Parents 3 April 1918.
61 Emmons, "Personal Goals, Life Meaning and Virtue: Wellsprings of a Positive Life", 105.
62 IWM PP/MCR/169 Private Papers Lt. F.A. Brettell, Letter from Peggy 29 December 1917.
63 Emmons, "Personal Goals, Life Meaning and Virtue: Wellsprings of a Positive Life", 105.
64 IWM 08/66/1 2nd Lt. D. Henrick Jones, Letters 25 December 1916 and 21 March 1918.
65 Liddle/WW1/GS/0144 Papers of A & C Black and Company.
66 IWM 07/02/1 Private Papers of S.B. Smith, Letters September–October 1916.
67 Szpunar, Watson, and McDermott, "Neural Substrates of Envisioning the Future".
68 Atance and O'Neill, "Episodic Future Thinking".
69 King, "The Health Benefits of Writing about Life Goals", 798–807; Pennebaker, "Writing about Emotional Experiences as a Therapeutic Process", 162–166.
70 IWM 98/28/1 Pte. F.G. Senyard, Letter to Wife 18 September 1917; RFM. ARC.3032 L/Cpl. C. White, 31 January 1917.
71 Pennell, *A Kingdom United*, 223.
72 IWM P 229 HE Trevor HET/5 Capt. Ashburner to Mrs Parker, Letter 14 October 1914.
73 IWM 96/24/1 W.M. Anderson, Letter to Wife 17 October 1916.
74 IWM 84/46/1 Hardie, "Report on III Army Morale, August–October 1917".
75 IWM 66/117/1 Private Papers of Capt. G.H. Greenwell, Letter 4 November 1916.
76 IWM 66/117/1 Private Papers of Capt. G.H. Greenwell, Letter 30 August 1916.
77 IWM Con Shelf Private Papers of Lt. R.D. Sheffield, Letter to Father 9 November 1914.
78 Snyder, Rand, King, Feldman, and Woodward, "'False' Hope".
79 Snyder, Irving, and Anderson, "Hope and Health: Measuring the Will and Ways", 287.
80 Snyder, Rand, King, Feldman, and Woodward, "'False' Hope", 1007.
81 Snyder, Rand, King, Feldman, and Woodward, "'False' Hope", 1007.
82 IWM PP/MCR/144 Private Papers of S. Judd, Diary 27 December 1914; IWM Con Shelf Private Papers of A.P. Burke, Letter to Reg 7 July 1916; IWM 06/5/2 Brig. Gen. G.A. Stevens, Letter to Mother Easter Monday 1918; Liddle/WW1/GS/0273 Papers of Capt. C. Carrington, Letter to Mother 28 March 1918 (no. 134).
83 Snyder, Rand, King, Feldman, and Woodward, "'False' Hope".
84 Liddle/WW1/GS/0273 Capt. C. Carrington, Letter to Mother 7 July 1918 and 13 October 1918.
85 IWM P430 Private Papers of Canon H.R. Bate, Memoir, 3; IWM 75/78/1 Private Papers of L/Cpl. K.M. Gaunt, Letter to 25 December 1914; IWM 96/24/1 Pte.

W.M. Anderson, Letters to Wife 1 and 6 December 1916; IWM 77/33/1 Lt. J.H. Johnson, Diary 23 December 1917 and 6 January 1918.

86 IWM 07/56/1 Private Papers of A.W. Andrews, Diary [written in 1917], 112–113.

87 IWM 90/17/1 Private Papers of A.G. Osborn, Diary 6 January 1915.

88 Brophy and Partridge, *The Long Trail: Soldiers' Songs and Slang 1914–18*, 126.

89 Brophy and Partridge, *The Long Trail: Soldiers' Songs and Slang 1914–18*, 126.

90 IWM Con Shelf A.P. Burke, Letter to Reg 29 December 1916.

91 Watson, *Enduring the Great War*, 182.

92 Hage, "A Not So Multi-Sited Ethnography of a Not So Imagined Community", 471.

93 IWM 08/94/1 Private Papers of 2nd Lt. W.J. Lidsey, Diary 3 November 1916.

94 IWM 05/8/1 Private Papers of W. Summers, Memoir, 16; "Advert: DO NOT READ THIS! Unless You Have a Girl at Home", *The B.E.F. Times*, No. 5, Vol. 2 (1918).

95 The National Archives WO 256/14 Diary of Field Marshal Haig, 1 December 1916.

96 IWM 06/5/2 G.A. Stevens, Letter to Mother Easter Sunday 1918; SOFO Box 16 Item 30 3/4/J3/9 C.T. O'Neill, Diary 25 March 1918; Liddle/WW1/GS/0266 Papers of Pte. E.A. Cannon, Poem from June 1918.

97 Snyder, Rand, King, Feldman, and Woodward, "'False' Hope", 1016.

98 Watson, *Enduring the Great War*, 11–43, 140–184.

99 Boniface, *Histoire religieuse de la Grande Guerre*; Houlihan, *Catholicism and the Great War*.

100 MR 3/17/110 Papers of Pte. H. Oldfield, Letter from HSS Clarke 10 CCS BEF to Mrs Oldfield on 20 November 1917; RFM.ARC.3032 L/Cpl. C. White, Diary 17 January 1917; IWM 96/24/1 W.M. Anderson, Letter to Wife 21 December 1916.

101 Gregory, *The Last Great War: British Society and the First World War*, 152–186.

102 This was particularly evident during 1916 in The National Archives WO 95/2085/3–4: War Diary 10th Battalion Royal Warwickshire Regiment.

103 Madigan, *Faith under Fire: Anglican Chaplains and the Great War*, 139–148.

104 IWM PP/MCR/269 Rev. E.N. Mellish, Letter in St Paul's, Deptford, Parish Church Magazine (March 1917) [Discussing Winter 1916].

105 Fussell, *The Great War and Modern Memory*, 312.

106 See IWM EPH 10150 "Ring" or IWM EPH 1915 "Bullet Crucifix". The latter was produced on the Italian Front but it is very likely that similar things were crafted in France and Belgium.

107 IWM 09/34/1 Private Papers of Capt. A.J. Lord, Letter 8 January 1918.

108 IWM 96/24/1 Pte. W.M. Anderson, Letter 21 December 1916.

109 IWM 96/24/1 Pte. W.M. Anderson, Letter 1 December 1916; RFM.ARC.2012.958 Ernest T. Marler, Diary 29 October 1916.

110 Webb, "Christian Hope and the Politics of Utopia", 113.

111 GWA (http://ww1lit.nsms.ox.ac.uk/ww1lit/items/show/5896, accessed 6 October 2016): Postcard Kept Throughout the War by Sapper E. Grantham; MR 3/17/110 Pte H. Oldfield, Postcards "The Divine Comforter" and "The White Comforter"; MR 3/17/139: Papers of H. Bridge, Postcard from Niece.

112 MR 3/17/110 Pte. H. Oldfield, Letter from HSS Clarke 10 CCS BEF to Mrs Oldfield on 20 November 1917; IWM 96/24/1 Pte. W.M. Anderson, Letter to Wife 21 December 1916.

113 IWM 03/30/1 Private Papers of W. Vernon, Letter 22 July 1918.

114 Gibson, *Behind the Front: British Soldiers and French Civilians, 1914–1918*, 383–384.

115 IWM PP/MCR/33 Private Papers of Maj. J.S. Knyvett, Diary 22 October 1914.

116 IWM 80/19/1 Private Papers of R.G. Plint, "If All the Rumours Came True", *The Banker's Draft*, No. 2, Vol 1 (July, 1916).

117 IWM 13/09/1 Private Papers of H. Milner, Memoir, 32.

118 IWM 76/27/3 Col. Hardman, Letter to Parents 4 April 1918.

119 IWM 77/83/1 Private Papers of S.A. Newman, Postcards of Arras.

120 RFM.ARC.2495 Sgt. S. Gill, Diary 20 December 1916.

121 IWM 06/5/2 G.A. Stevens, Letter to Mother 11 November 1916; British Library – “Editorial”, *The B.E.F. Times*, No. 3 Vol. 1 (1917), 1–2; British Library – “Editorial”, *The B.E.F. Times*, No. 5 Vol. 2 (1918), 1–2.
122 Liddle/WW1/GS/0313 Papers of Pte. C. Clarke, Memoir, 14; IWM 66/96/1 Private Papers of Reverend M.A. Bere, Letter to wife 17 September 1917.
123 Bairnsfather, *Best of Fragments from France*, 10.
124 Gibson, *Behind the Front: British Soldiers and French Civilians, 1914–1918*, 12.
125 IWM 09/34/1 A.J. Lord, Letter 1 April 1918; Jones, “The Psychology of Killing”, 244; Kestnbaum, “The Sociology of War and the Military”, 242–243.
126 Liddle/WW1/GS/0144 A & C Black Publishers, Letter from E. Lindsell 23 June 1917.
127 SOFO Box 16 3/4/N/2 Lt. Reginald N. Neville, Letter to Sister 2 August 1917.
128 Strachan, “Training, Morale and Modern War”, 216.
129 IWM 12/36/1 Private Papers of F. Hubard, Letter to Wife 25 October 1916.
130 Foucault, “What is an Author?”, 216.
131 IWM Con Shelf A.P. Burke, Letters to Reg 13 June and 12–15 October 1916.
132 IWM 08/66/1 2nd Lt. D. Henrick Jones, Letter 25–26 December 1916; MR 3/17/126 Pte. John Peat, Letter 11 October 1917; IWM 01/21/1 H.T. Madders, Diary 17 March 1918; RFM.ARC.2012.146.1 Albert Victor Arthur, Diary 20 September 1917.
133 IWM PP/MCR/169 F.A. Brettell, Letters from Peggy.
134 Hewitson, “‘I Witness’: Soldiers, Selfhood and Testimony in Modern Wars”, 313.
135 IWM Interview 554 C.R. Quinnell, Reel 15.
136 Neitzel and Welzer, *On Fighting, Killing and Dying*, 340.
137 IWM 76/27/3 Col. Hardman, Letter 3 April 1918; IWM 77/33/1 Lt. J.H. Johnson, Diary 28 December 1917; IWM Misc 108 Item 1710 Pocket Diary, 7 September 1914.
138 Audoin-Rouzeau and Becker, *14–18: Understanding the Great War*, 24.
139 IWM P 146 Private Papers of Lt. Col. C.S. Baines, Memoir p. 7.
140 IWM P 146 Private Papers of Lt. Col. C.S. Baines, Memoir p. 7.
141 IWM 77/33/1 Lt. J.H. Johnson, Diary 6 November 1918.
142 SOFO Box 23 Item 99 Papers of Capt. James Harold Early, Diary 4 May 1915; IWM 77/33/1 Lt. J.H. Johnson, Diary 28 December 1916; IWM 06/5/2 Brig. General G.A. Stevens, Letter to Mother 11 November 1916; IWM 77/33/1 Lt. J. H. Johnson, Diary 28 December 1917.

Bibliography

Archival material

British Library
The B.E.F. Times, No. 2, Vol. 1 (1916)
The B.E.F. Times, No. 3, Vol. 1 (1917)
The B.E.F. Times, No. 5, Vol. 2. (1918)
Cambridgeshire Territorial Gazette, No. 4(1916)

Great War Archive (GWA), Oxford University [Online Database]
GWA (*http://ww1lit.*nsms.ox.ac.uk/ww1lit/items/show/5896, accessed 6 October 2016): Postcard Kept Throughout the War by Sapper E. Grantham

Imperial War Museum, London
IWM 01/21/1: Private Papers of H.T. Madders
IWM 02/43/1: Private Papers of R.E.P. Stevens
IWM 03/30/1: Private Papers of W. Vernon

IWM 05/8/1: Private Papers of W. Summers
IWM 06/5/2: Private Papers of Brig. General G.A. Stevens
IWM 07/02/1: Private Papers of S.B. Smith
IWM 07/56/1: Private Papers of A.W. Andrews
IWM 08/66/1: Private Papers of 2nd Lt. D. Henrick Jones
IWM 08/94/1: Private Papers of 2nd Lt. W.J. Lidsey
IWM 09/34/1: Private Papers of Capt. A.J. Lord
IWM 12/36/1: Private Papers of F. Hubard
IWM 13/09/1: Private Papers of H. Milner
IWM 66/96/1: Private Papers of Reverend M.A. Bere
IWM 66/117/1: Private Papers of Capt. G.H. Greenwell
IWM 75/78/1: Private Papers of L/Cpl. K.M. Gaunt
IWM 76/27/3: Private Papers of Col. Hardman
IWM 77/33/1: Private Papers of Lt. J.H. Johnson
IWM 77/83/1: Private Papers of S.A. Newman
IWM 80/19/1: Private Papers of R.G. Plint
IWM 84/46/1: Private Papers of Capt. M. Hardie
IWM 90/17/1: Private Papers of A.G. Osborn
IWM 96/24/1: Private Papers of Pte. W.M. Anderson
IWM 98/28/1: Private Papers of Pte. F.G. Senyard
IWM 99/13/1: Private Papers of Major S.O.B. Richardson
IWM Con Shelf: Private Papers of A.P. Burke
IWM Con Shelf: Private Papers of Lt. R.D. Sheffield
IWM Misc 108 Item 1710: Pocket Diary
IWM P 146: Private Papers of Lt. Col. C.S. Baines
IWM P 229: Private Papers of Brig. Gen. H.E. Trevor
IWM P 430: Private Papers of Canon H.R. Bate
IWM PP/MCR/33: Private Papers of Maj. J.S. Knyvett
IWM PP/MCR/144: Private Papers of S. Judd
IWM PP/MCR/169: Private Papers Lt. F.A. Brettell
IWM PP/MCR/269: Private Papers of Reverend E.N. Mellish
IWM EPH 1915: "Bullet Crucifix" (artefact)
IWM EPH 10150: "Ring" (artefact)
IWM Interview 554: C.R. Quinnell

Liddle Collection, University of Leeds
Liddle/WW1/GS/0144: Papers of A & C Black and Company
Liddle/WW1/GS/0266: Papers of Pte. E.A. Cannon
Liddle/WW1/GS/0273: Papers of Capt. C. Carrington
Liddle/WW1/GS/0313: Papers of Pte. C. Clarke
Liddle/WW1/GS/0583: Papers of S. Frankenburg
Liddle/WW1/GS/0746: Papers of H.O. Hendry

Manchester Regiment (MR) Archive, Ashton-Under-Lyne
MR 2/17/65: Papers of John Douglas Powell
MR 3/17/110: Papers of Pte. H. Oldfield
MR 3/17/126: Papers of Pte. John Peat
MR 3/17/139: Papers of H. Bridge

The National Archives, Kew
WO 95/2085/3–4: War Diary 10th Battalion Royal Warwickshire Regiment
WO 256/14: Diary of Field Marshal Haig
National Army Museum (NAM), London
NAM 1959-03-34(1): *The B.E.F. Times*, No. 2 Vol. 1 (Monday 25 December 1916)
NAM 1959-03-34(4): *The Somme Times: with which are incorporated The Wipers Times, The "New Church" Times & The Kemmel Times*, No. 1 Vol. 1 (Monday 31 July 1916)
NAM 1992-09-139-6: Papers of A. Joy
NAM 2005–2–6: Papers of Capt. Maurice Asprey
NAM 7403-29-486-144: Papers of Sgt. Harry Hopwood

Royal Fusiliers (RFM) Archive, London
RFM.ARC. 2012.146.1: Albert Victor Arthur
RFM.ARC. 2012.958: Ernest T. Marler
RFM.ARC. 2495: Sgt. S. Gill
RFM.ARC 2013.8.2: Pte. P.N. Wright
RFM.ARC. 3032: L/Cpl. C. White

Soldiers of Oxfordshire (SOFO) Museum Archive, Woodstock
Box 16 3/4/N/2: Lt. ReginaldN. Neville
Box 16 Item 29 3/4/J3/8: Lt. J.E.H. Neville
Box 16 Item 30 3/4/J3/9: C.T. O'Neill
Box 16 Item 61: Collection of 15 Patriotic Cards 1914–1918Vintage
Box 23 Item 99: Capt. James Harold Early

Published material

Atance, C.M. and D.K. O'Neill. "Episodic Future Thinking." *Trends in CognitiveSciences*, 5, no. 12(2001): 533–539.

Audoin-Rouzeau, S. *Men at War 1914–1918: National Sentiment and Trench Journalism in France during the First World War.* Translated by H. McPhail. Oxford: Berg Publishers, [1992] 1995.

Audoin-Rouzeau, S. and A. Becker. *14–18: Understanding the Great War.* Translated by C. Temerson. New York: Hill & Wang, 2002.

Bairnsfather, B. *Best of Fragments from France.* Edited by T. Holt and V. Holt. Barnsley: Pen and Sword, [1978] 2014.

Becker, J.J. *The Great War and the French People.* London: Bloomsbury Academic, 1986.

Beckett, I., T. Bowman, and M. Connelly. *The British Army and the First World War.* Cambridge: Cambridge University Press, 2017.

Boff, J. "The Morale Maze: The German Army in Late 1918." *The Journal of Strategic Studies* 37, no. 6–7(2014): 855–878.

Boff, J. *Winning and Losing on the Western Front: The British Third Army and the Defeat of Germany in 1918.* Cambridge: Cambridge University Press, 2012.

Bowman, T. *The Irish Regiments in the Great War: Discipline and Morale.* Manchester: Manchester University Press, 2004.

Boniface, X. *Histoire religieuse de la Grande Guerre*, Paris: Fayard, 2014.

BrophyJ. and E. Partridge. *The Long Trail: Soldiers' Songs and Slang 1914–18.* London: Sphere, 1969.

Brotemarkle, R.A. "Development of Military Morale in a Democracy." *Annals of theAmerican Academy of Political and Social Science*, 216(1941): 79–87.

Brown, C., T. Nardin, and N. Rengger (eds). *International Relations in PoliticalThought: Texts from the Ancient Greeks to the First World War*. Oxford: Oxford University Press, 2005.

Bullock, A. and O. Stallybrass. *The Fontana Dictionary of Modern Thought*. London: HarperCollins, 1977.

Carver, C.S. and M. Scheier. "Optimism." In *Positive Psychological Assessment: A Handbook of Models and Measures*, edited by S.J. Lopez and C.R. Snyder, 75–89. Washington, DC: American Psychological Association, 2003.

Emmons, R.A. "Personal Goals, Life Meaning and Virtue: Wellsprings of a Positive Life." In *Flourishing: Positive Psychology and the Life Well Lived*, edited by C.L.M. Keyes and J. Haidt, 105–128. Washington, DC: American Psychological Association, 2003.

Englander, D. "Discipline and Morale in the British Army, 1917–1918." In *State, Society and Mobilization in Europe during the First World War*, edited by J. Horne, 125–143. Cambridge: Cambridge University Press, 1997.

Fennell, J. *Combat and Morale in the North African Campaign*. Cambridge: Cambridge University Press, 2011.

Ferguson, N. *The Pity of War: Explaining World War I*. London: Penguin, 1998.

Fletcher, A. *Life, Death and Growing Up on the Western Front*. Cambridge, MA: Yale University Press, 2013.

Foucault, M. "What is an Author?" In *Aesthetics, Methods and Epistemology*, edited by J. D. Faubion and translated by R. Hurley*et al.*, 205–222. New York: The New Press, 1998.

Frederickson, B.L. "Why Choose Hope?" *Psychology Today*. 23 March 2009. www.psychologytoday.com/blog/positivity/200903/why-choose-hope.

Fuller, J.G. *Troop Morale and Popular Culture in the British and Dominion Armies, 1914–1918*. Oxford: Oxford University Press, [1991] 2001.

Fussell, P. *The Great War and Modern Memory*. Oxford: Oxford University Press, [1977] 2000.

Gibson, C. *Behind the Front: British Soldiers and French Civilians, 1914–1918*. Cambridge: Cambridge University Press, 2014.

Gregory, A. *The Last Great War: British Society and the First World War*. Cambridge: Cambridge University Press, 2008.

Hage, G. "A Not So Multi-Sited Ethnography of a Not So Imagined Community." *Anthropological Theory* 5, no. 4(2005): 463–475.

Hewitson, "'I Witness': Soldiers, Selfhood and Testimony in Modern Wars." *German History* 28, no. 3(2010): 310–325.

Horne, J. "Entre expérience et mémoire: Les soldats français de la Grande Guerre." *Annales. Histoire, Sciences sociales* 60 (2005): 903–919.

Houlihan, P. *Catholicism and the Great War: Religion and Everyday Life in Germany and Austria-Hungary, 1914–1922*. Cambridge: Cambridge University Press, 2015.

Jones, E. "The Psychology of Killing: The Combat Experience of British Soldiers during the First World War." *Journal of Contemporary History* 41, no. 2(2006): 229–246.

Kestnbaum, M. "The Sociology of War and the Military." *Annual Review of Sociology* 35 (2009): 235–254.

King, L.A. "The Health Benefits of Writing about Life Goals." *Personality and Social Psychology Bulletin* 27, no. 7(2001): 798–807.

Kitchen, J. *The British Imperial Army in the Middle East: Morale and Identity in the Sinai and Palestine Campaigns*. London: Bloomsbury, 2014.

Lopez, S.J., C.R. Snyder, and J.T. Pedrotti. "Hope: Many Definitions, Many Measures." In *Positive Psychological Assessment: A Handbook of Models and Measures*, edited by S.J. Lopez and C.R. Snyder, 91–107. Washington, DC: American Psychological Association, 2003.

Madigan, E. *Faith under Fire: Anglican Chaplains and the Great War*, London: Palgrave Macmillan, 2011.

Marshall, S.L.A. *Men Against Fire: The Problem of Battle Command in Future War*. Norman: Oklahoma University Press, [1947] 2000.

McCartney, H. *Citizen Soldiers: The Liverpool Territorials in the First World War*. Cambridge: Cambridge University Press, 2005.

Neitzel, S. and H. Welzer. *On Fighting, Killing and Dying: The Secret Second World War Tapes of German POWs*. New York: Alfred A. Knopf, 2012.

Pennebaker, J.W. "Writing about Emotional Experiences as a Therapeutic Process." *Psychological Science* 8, no. 3(1997): 162–166.

Pennell, C. *A Kingdom United: Popular Responses to the Outbreak of the First World War in Britain and Ireland*. Oxford: Oxford University Press, 2012.

Reivich, K. and J. Gillham. "Learned Optimism: The Measurement of Explanatory Style." In *Positive Psychological Assessment: A Handbook of Models and Measures*, edited by S.J. Lopez and C.R. Snyder, 57–74. Washington, DC: American Psychological Association, 2003.

Roper, M. *The Secret Battle: Emotional Survival in the Great War*. Manchester: Manchester University Press, 2009.

Roper, M. "Nostalgia as an Emotional Experience in the Great War". *The Historical Journal* 54, no. 2(June, 2011), 421-451.

Sheffield, G. *Command and Morale: The British Army on the Western Front 1914–1918*. Barnsley: Pen & Sword, 2014.

Shephard, B. *A War of Nerves: Soldiers and Psychiatrists 1914–1994*. London: Pimlico, 2002.

Snyder, C.R., L. Irving, and J.R. Anderson. "Hope and Health: Measuring the Will and Ways." In *Handbook of Social and Clinical Psychology: The Health Perspective*, edited by C.R. Snyder and D.R. Forsyth, 285–305. Elmsford: Pergamon Press, 1991.

Snyder, C.R., K.L. Rand, E.A. King, D.B. Feldman, and J.T. Woodward. "'False' Hope." *Journal of Clinical Psychology* 58, no. 9(2002): 1003–1022.

Stevenson, D. *With Our Backs to the Wall: Victory and Defeat in 1918*. London: Penguin, 2014.

Strachan, H. "Training, Morale and Modern War." *Journal of Contemporary History* 41, no. 2(2006): 211–227.

Szpunar, K.K., J.M. Watson, and K.B. McDermott. "Neural Substrates of Envisioning the Future." *Proceedings of the National Academy of Sciences of the United States of America* 104, no. 2(2006): 642–647.

Tholas-Disset, C., and K.A. Ritzenhoff. *Humor, Entertainment, and Popular Culture during World War I*. New York: Palgrave Macmillan, 2015.

Watson, A. *Enduring the Great War: Combat, Morale and Collapse in the German and British Armies, 1914–1918*, Cambridge: Cambridge University Press, 2008.

Watson, A. "Morale." In *The Cambridge History of the First World War: Volume II, The State*, edited by Jay Winter, 174–195. Cambridge: Cambridge University Press, 2014.

Webb, D. "Christian Hope and the Politics of Utopia." *Utopian Studies* 19, no. 1 (2008): 113–144.

Wilcox, V. "'Weeping Tears of Blood': Exploring Italian Soldiers' Emotions in the First World War." *Modern Italy* 17, no. 2(2012): 171–184.

Wilcox, V. *Morale and the Italian Army during the First World War.* Cambridge: Cambridge University Press, 2016.

Index

For Product Safety Concerns and Information please contact our EU
representative GPSR@taylorandfrancis.com
Taylor & Francis Verlag GmbH, Kaufingerstraße 24, 80331 München, Germany

www.ingramcontent.com/pod-product-compliance
Ingram Content Group UK Ltd.
Pitfield, Milton Keynes, MK11 3LW, UK
UKHW021302190926
475701UK00009B/589
9780367588922